ASTROGENETICS

An Interdisciplinary Theme of Natural Control System (of Animates & Inanimates)

ASTROGENETICS

An Interdisciplinary Theme of Natural Control System (of Animates & Inanimates)

C. N. Jha
Ph-D Retired, Principal C.M. Sc. College,
University Professor and Head L. N. Mithila,
University, Darbhanga, Bihar, India

NAMYA PRESS.com

Published by

Namya Press

India: 213, Vardaan House, 7/28 Ansari Road, Daryaganj, Delhi- 110002
USA: 10685-B Hazlehurst Dr. # 31013 Houston, Texas - 77043
Email: namyapress@gmail.com
Website: https://namyapress.com

Edition	:	First Published in 2023 (Namya Press)
Title	:	Astrogenetics
Author	:	C.N. Jha
ISBN	:	978-93-5545-009-8

Preface

We are all surrounded by high energy particles coming from Sun, Stars and galactic nabulae but Modern Life Science ignores it.- HOW ? ? ?

1. Cosmic rays coming from SUN, STARS and GALACTIC NEBULAE (360°)
2. Cosmic rays interact among themselves
3. They interact with large chemical surfaces of planets.
4. Most of the rays are absorbed in ionosphere (ozone layer etc.)
5. Few of them are reflected by earth's gravitational force.
6. Rest cosmic rays enter into earth's atmosphere.
7. Cosmic rays are high energy particles of Shorter, Wave length and variable intensities (1 to 1 million Bevestron).
8. Each particle reacts with earth's environment producing gigantic shower of rays thus covering whole surface of the earth within and across (so, the concept in Adhyatma prevails as kan-kan me vyapt).
9. Rays hit individuals in different angles, power and interactions in three different layers.
10. At low intensity hit body physiology but at higher intensity effect internal parts of the body and at greatest intensity for longer duration effect Genetic materials.
11. The rays of high intensity for Longer duration only cause mutation. They also form habit of the individuals. Bad

rays may cause Genetic diseases, heritable and non-heritable changes.

12. Rays are beneficial and adverse, hot and cold. The bad rays of different durations lay foundation of different diseases. Then only foreign agents (bacteria, virus etc) effect. Thus foreign agents are secondary causes, Ray primary.
13. Bad rays - Rays contributed by stars and others interact with the surface of the Planets and then hit individuals in bad angles. Rays coming from stars of the characteristics of Rahu, Saturn, Ketu, Sometimes, Mars, also Sun are bad/or even rays coming from good characteristics of stars and interact with above noted bad planets may be bad. This is the key of the Mechanism. Details have been given with illustrations and evidences are expressed. But on the basis of discussions and illustration in both the chapters it is crystal clear - that cosmic Rays are causes of all human diseases and Genetic Phenomenon.

The ideas derived are not merely hypothesis rather based on self experiments on atleast 20 thousand patients of various ailments from almost every strata and walks of life under a voluntary organization (Astrogenetic Society, Kathalwari, Darbhanga, Bihar, India), following the guidelines of ICMR, New Delhi for handling of Human patients on the principle of "effect to cause" analysis because the causes are remote. Foundation of ideas derived inter-disciplinary from old science Astrology, Astrophysics, life science, chemistry, physics, Basic tenets of medical science etc. Discussions carried over with experts of such fields at occasions on the plateform of the Society during handling of patients. Records of patients were maintained with etiology in about 36 volumes but started after 5 years of inception due to gradual confirmation of thoughts/experiments (development of dozes and durations). Patients expressions reflected through print media were already enclosed. During exploration and rendering services to ailing mankind, regular communications were conducted with high scientific institutes of the world (enclosed).

Ultimately this is my utmost feeling at the threshold of activities mentioned, the theme derived regarding Natural control system of animates and inanimate thus producing every event must be conceived upon by all who stepped in this cosmos. There is a magnificient convergence of science and Adhyatma that ever for oblivious but not for abdurate, through scientific derivations. This will certainly help to encourage of Humanity.

The Therapeutic part is by product of the idea but most useful for mankind in releasing himself and society from ailments. where career may be set upon.

Moreover, exogenous applications of RAD's their instant reflections as effects (in tools) within ailing patients, tell some extramural facts neverthless to be ignored. It further extricate a vital fact that all activities are extraterritorial on earth crust, even wealth of every penny. Of course, it is hard to believe this true exposition.

Some vital discoveries during **exploration** are aptly conveyed to the contemporary Scientific world. This vast **exploration** (of 25 yrs) while **rendering** services, crystallized few bewildering facts of Nature, in contrary to established theories of modern science, extracted from inner fabrics of magnificient world/ Nature. During escapade, it all gathered that live "Nature" exhibits all genetic activities within realistic Cosmographs of individuals computed after interactions, emancipating lively genetic activities apparent in pedigree charts perhaps in every family.

Hither to, various underlined scientific discoveries have already been revised on this platform (with evidences) from contemporary views regareding- Sun Rays, Gravitational forces, Aura, Natural control system, relationship of living with living and living with non living to facilitate "Events" in cosmos/and on planet earth. Though the fact uttered is an exhort to Scientific world to reconcile after due attention on this expendient, Since all have experimental evidences. However, it is turning point with vast approach and will be a rescue from many detrimental effects for the benefit in world Scenario especially to Living-Non Living.

Letters from High Scientific Institutes

SCIENCE COUNCIL OF JAPAN
CONSEIL DES SCIENCES DU JAPON

22-34, Roppongi 7 chome, Minato-ku, Tokyo 106 Japan
Telephone: 03-403-6291
Cable address: SCIENCOUNCIL TOKYO

January 9, 1989

Dr. C.N.Jha
Readers in Botany
C.M.Sc College
Darbhanga, North Bihar

Dear Dr. Jha

We've heard about your "Astro Genetics" from Dr. Jiro Kondo, president, Science Council of Japan.
However, we're sorry to inform you our committees are not in the position to commit ourselves to doing something on your research.

We hope your research will proceed well.

With best regards,

Sincerely yours,

Yukio Hiramoto

Yukio Hiramoto
chairman of Research Liaison
Committee of Zoology

Ikuo Takeuchi

Ikuo Takeuchi
chairman of Research Liaison
Committee of Botany

All India Institute of Medical Sciences
Genetics Unit, Department of Pediatrics
Old Operation Theatre Building
Ansari Nagar, New Delhi-110 029, INDIA

Ishwar, C. Verma, F.R.C.P. (Lond). F.A.M.S. F.A.A.P. (USA)
Professor of Pediatrics
Officer-in-Charge, Genetics Unit
Editor-in-Chief, The Indian Journal of Pediatrics

W.H.O. Collaborating Centre For Training
In Clinical and Laboratory Genetics
In Developing Countries

In reply please quote Ref. No.

Mr. C.N. Jha
Founder and Director
Astrogenetic Society
Kathalwari, Darbhanga

October 20, 1993

Dear Mr. Jha,

This is in response to your letter dated 19.9.93. We read with interest your experience with cosmic rays therapy. We at present are totally ignorant about this issue. Before planning any collaborative project we want to meet you and discuss your project plans. You may plan to come to our unit sometimes in December, 1993 or January 1994 for 1-2 days. Please do write to us before finalising your programme.

With best wishes,

Sincerely,

I.C. VERMA
Prfessor of Pediatrics

TEL. : 667136 6963980
6962794 6962895

तार : विज्ञानी
GRAM : SCIENTIFIC
टेलेक्स/TELEX. : 031-73067
फैक्स/FAX : 011-6868662

भारतीय आयुर्विज्ञान अनुसंधान परिषद
INDIAN COUNCIL OF MEDICAL RESEARCH
अन्सारी नगर, पोस्ट बॉक्स 4911, नई दिल्ली–110029.
ANSARI NAGAR. POST BOX 4911. NEW DELHI-110029

Regd

No. 16/3/92-Admn. II(Pt.II) Dated : 8.2.1996
15

To

Dr.C.N. Jha,
Univ. Prof. in Botany,
C.M.Sc. College,
Darbhanga,
BIHAR.

Subject : Astrogenetics (a 'sample' study before and after Astrogenetic Regulations)

......

Sir,

Reference your letter No. AG-100114/95 dated 11.12.1995. I am directed to request you to kindly submit a Research Proposal on the above mentioned subject for consideration of an extramural funding.

An application form for grant-in-aid is sent herewith. Further necessary action will be taken on receipt of the Project Proposal by the Council.

Yours faithfully,

(S.B. Tewari)
Administrative Officer
for Director-General.

Copy for information to Dr.I.C. Verma, Prof. of Pediatrics, Genetics Unit, Deptt. of Pediatrics, Old Operation Theatre Building, AIIMS, New Delhi alongwith a copy of the letter under reference.

(S.B. Tewari)
Administrative Officer
for Director-General

भारतीय प्रौद्योगिकी संस्थान मुंबई
पवई, मुंबई-400 076, भारत

Indian Institute of Technology Bombay
Powai, Mumbai-400 076, India

Phone : (022) 578 2545, 578 6530
Gram : Technology, Mumbai-76
Telex : 011-72313 IITB IN
Fax : 091-22-578 3480

IIT Bombay

Prof. S.Durani
Convener
Biotechnology Centre

12th Nov. 1997

Prof. C.N.Jha
Department of Botany
CM Science College
Darbangha, Bihar

Dear Prof. Jha,

This has reference to our discussion on your interest in genetic analysis using modern tools provided by molecular biology. Several of my colleagues are with the expertise of relevant kind and could in principle help you based on mutual interest. Our academic semester is now coming to an end and our faculty are likely to have some free time available for consultations and discussions. Hence I would like to advise that you could visit our Centre during the fortnight Nov. 21 to Dec. 7 to pursue the necessary discussions.

With kind regards,

Yours sincerely,

S Durani

(S.Durani)

Contents

1

Mechanism

Interaction of cosmic rays with body physiology and genetic materials is called astrogenetics.

Then what are cosmic rays?

Rays come from the Sun, stars and galactic nebulae. They are definite scientific entities. It is never fantastic or imaginary as depicted in myths. So we are all surrounded by high energy particles, i.e., cosmic rays, but modern life science ignores. How is it possible? Cosmic rays are coming from 360°, i.e., all around the earth. So they interact among themselves, further with large chemical/mineral surfaces of planets. Few of them are then absorbed in the ozone layer (inosphere). Then a few are rejected by the earth's gravitational force. Still remaining cosmic rays enter into the earth's atmosphere. These cosmic rays thus available on the earth's surface are the most powerful energy particles that are managing and maintaining all other types of energies. They are of a shorter wavelength and with variable intensities, i.e., 1 to 1 million Bevestron. Each particle reacts with the earth's environment producing gigantic energy. The shower of rays (energy) covers the whole surface of the earth around and across. This is the concept described in religious books as omnipresent, super power (kan-kan me vyapt). These rays maneuvour all sorts of activities. So, this is the fundamental basis of modern science and *Adhyatma* as well.

Consequently no one can find even a pin head space on earth, free from cosmic rays. Pertaining to the fact any experiment on cosmic rays cannot be carried out in total deficient surface on earth rather it is only possible on increased conditions.

Rays hit individuals at different angles, power and interaction, in three different layers. At low intensity, they affect body physiology but at higher intensity they affect internal parts of the body, and at the highest intensity they affect genetic materials. They cause mutation rarely.

Different actions of the rays result differently, as follows:

1. At low intensity but in broad spectrum they cause minor ailments by manifesting unfavorable environmental factors. People go for self-treatment and rarely take doctors' advice.
2. Higher intensity of the rays for some longer duration affect internal organ, causing established diseases.
3. At the highest intensity for longer duration, the rays affect genetic materials. It includes the following:
    a. Rays of the highest intensity for longest duration, cause mutation, only when other forces are weak.
    b. Rays affecting genetic material develop nature and habit of the individuals.
    c. Bad rays may cause genetic diseases either heritable changes or non-heritable changes.
    d. Unlike our prevailing knowledge, rays are also beneficial, and adverse, hot, and cold. The bad rays of different durations lay foundation of diseases. Then only a foreign agents such as bacteria, virus, etc. affects others. Thus, they are called secondary causes.

It has been observed that cosmic rays manage and maintain electromagnetic forces so percepted during different activities.

Thus, on the basis of pretext, and experiment carried over, it may be suspected that 'cosmic rays are causes of all human diseases and genetic phenomenon' after a discussion with Genetic Dept., AIIMS (through several correspondences and personal meeting.).

Three layered effects of the rays (as discussed) appear covering all activities of individuals. In the first two layers rays are responsible for wellness and diseases. But at high intensity rays effect the third layer, i.e., genetic materials specifically manupulating gene action, might be deciding Nature and Fate of the individuals.

This is based on the experiments and studies of cosmography of over 15000 patients and individuals.

Going through these tough discussions, people might be feeling awkward, but every assertion is based on experiments with 15-20 thousand people within 25 years. The remarkable facts were placed before high scientific institutes like AIIMS, New Delhi and Science Council, Japan for scrutiny. AIIMS, New Delhi discussed first through communications and then called for a personal meeting. Scientific papers correspondences published.

Science Council, Japan, called meeting of the Liaison committee and declared the research good but new to science. It cheered and wished to continue. ICMR requested to submit a project on the topic it assigned in collaboration with AIIMS. But unfortunately due to the paucity of time, since running a voluntary organization and rendering services at the same time, failed to comply with. The then V.C. sanctioned a leave to participate in the discussion with AIIMS. He also asked the requirements of astrogenetics, and I submitted the required papers. But the Manav Kalyan Forum was running in full swing in the society so failed to pursue the files in office. Anyway, society always remained pure voluntary (Manav Kalyan forum) without any fee, subscription and aid. Fellow persons were purely rendering free services. So, they couldn't stay for long in this voluntary society. 3-4 sets of persons participated. This remained the story of this voluntary organisation (society).

Ultimately the astrogenetic society had to be closed after rendering 25 years of active services to mankind. However, it will never be closed completely as patients percepted truth and power of the therapy.

The patients who were regulated for the whole life are still in contact even through phone for slight changes in regulations.

Few typical patients denied by medical organisations still manage to be regulated here.

It appears that cosmic rays support energy, to gene action hence there is concept of less functioning, functioning or not functioning genes. Cosmic rays, due to interactions, fall under various types. During interaction, they interact among themselves i.e. the star of whole sky 360° have been characterized after diving into 12 parts. Each part is known as Rashi (i.e. 30° of sky) considered as of different characteristics. So all parts are different due to different characteristics of stars. Again for minute characterization, the whole sky was divided into 27 parts named Constellations. Besides, there are 9 important planets. (Other planets discoverd, appeared more or less similar in characteristcs.) So, total interactions of rays come to 8503056 types of interactions.

The total interactions are as following:

= rashis x rashis x constellations x constellations x planets x planets

= 12 x 12 x 27 x 27 x 9 x 9

The basic feature obviously considered in natural phenomenon, is matter-energy interconversion and transformation in this universe i.e. in this living system, genetic material is matter. Where as cosmic rays are sources of energy. So everything in the living system is resultant of the reactions and transformations of only these two.

Cosmic rays do not only react with genetic materials of the living system rather they also react with nonliving materials.

The effect of cosmic rays on living bodies ultimately results in the aura formation around the body. This formation also takes place in the nonliving system. But the difference lies in proteinaceous colours. In the proteinaceous body, the Aura is multicoloured while it is monocoloured in nonliving systems.

Aura

"What is AURA?" is definitely a question. The electronic body which develops all around the main body after interactions is called AURA. It develops both in living and nonliving systems, as told earlier. The aura has positive and negative impulses such as attraction and repulsion forces in relation to others, depending upon the nature of protien and simultaneously, the nature and power of interacting cosmic rays. So **AURA** is an electronic reflection of an internal body, i.e., protein and all. Modern science has developed a special camera, and photography for this electronic body. This is known as **Kirlian** photography.

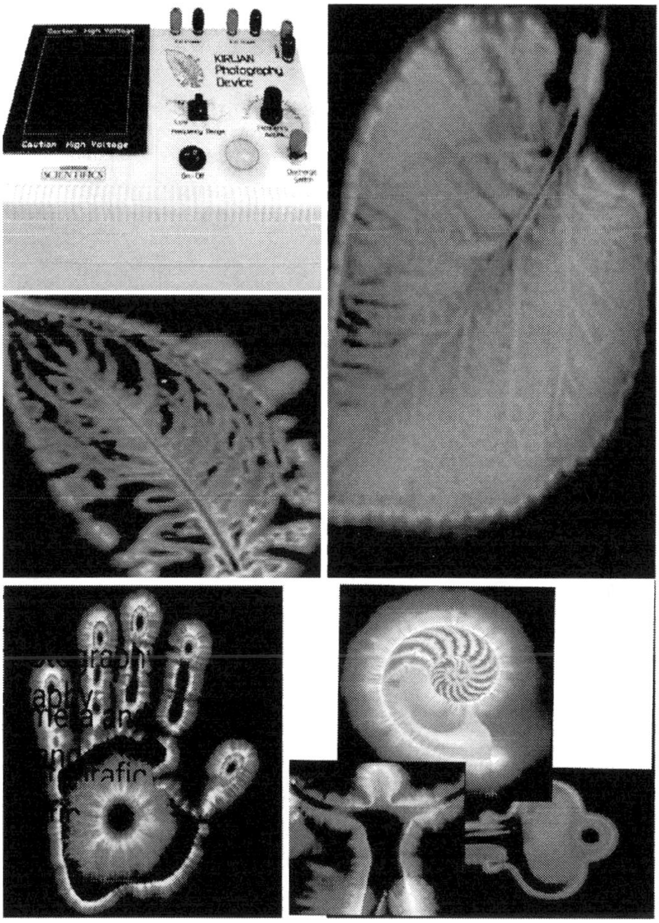

Fig. 1: Aura from Kirlian photography of living and non living objects.

This is for crime detection and disease diagnosis. Similarly an electronic body also appears around each nonliving (inanimate) but of a single colour. However, they have charged potentials.

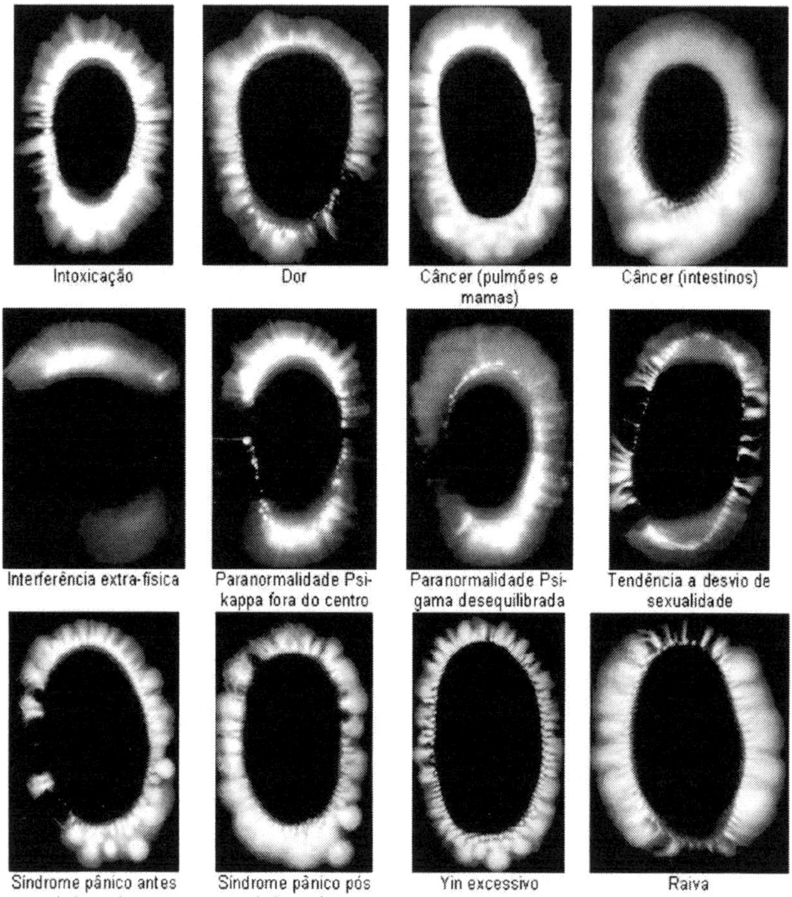

The aura mechanism is wonderful, presentation of nature in a very simple way. It produces attraction-repulsion mechanism which is the basic/core activity of the natural phenomenon and relationship mechanism. The whole solar system is balanced on this principle.

Each planet has a particular chemical/mineral surface under interaction with cosmic rays. So, they develop vast gravitational force acting on the same principle attraction and repulsion as in the tiny objects, i.e., animates and inanimates interacting and producing various effects (will be apparent in the further discussion). Thus the aura formation in every case develops magnetic force due to attraction and repulsion mechanism. This is the key mechanism of a relationship between living-living and living-nonliving.

This, in nutshell, explains the whole activity of nature both in relation to living and nonliving.

The aura of a person manifests other activities also by interacting with even nonliving or inanimates. A positive and powerful aura of a person when interacts with a powerful aura of nonliving, for example, building materials, it is easily procured and building is constructed without any hindrance.

However, this falls under a specific mechanism of aura, knows as the fate mechanism with living to living (all relation, friends and foe) and living to nonliving, fullfilling daily needs. Eventually they depend upon weak and powerful interactions. Weak interactions when become powerful, a tea vendor may develop as very big man.

So, the aura mechanism is also fate specific, and good and bad effects depend on the nature and power of interactions of cosmic rays in relations to the genetic material under interaction.

Fate is sometimes misnomer and debatable. Some people tell that it is depending on one's effort. But from where is the power of effort achieved? So, a simple truth is unveiled of aura mechanism for fate.

According to astrogenetics, fate is merely based on aura, since it depicts the mechanism of attraction-repulsion. So, it is depending upon the nature of genetic materials one inherits from parents and interaction of rays, developing in a different period of the individual. Thus it is specifically time dependent. Accordingly, friends become foe at particular juncture of time. A dragger's drown enemy becomes kith and kin. Same is the case with different relations, wife, lover, even distant relatives and social relationships. None of them is perpetual. Every type of relationship may fluctuate if adverse rays influence it. So, these are the different time specific effect of cosmic rays mediating and manipulating through aura. The power of rays with good interactions emancipates powerful good effects spreading to remote places like fame and vice-versa. It is well known that a person has a large number of followers but after a while all vanishes and the person hardly maintains himself. This is the periodic power of rays acted through aura. As the good rays is switched off aura changes, the same person becomes unacknowledged. These are the dramas of periodic rays or interactions acting through Aura formation. Everybody sees to it but hardly believes it. The dominating auras of big personalities supresses the weak Auras of surrounding. So the surroundingfirst becomes subdued, or over influenced. So, it was thoroughly outlined during study that nature remained very simple in the mechanism by developing the genetic system and aura formation (in living and nonliving). Marvellous simile for characterization, and organogenesis was obtained here. The clue resembles to a great extent, showing a thorough pattern. Natural resemblances are never exact it is simply symbolising. In that way, it reveals that there are 12 pairs of cranial nerves, and there are 12 important parts of sky, different from each other in stars characteristics Every part (30°) is called roughly *Rashi*. So there are 12 *rashies* (big parts for characterization, consisting of stars, as they are rays contributing). The pattern of organogenesis appears similar as the first pair of cranial nerve supplies to the same organs of the human body as the first *Rashi* indicates, so on and so forth. The full details of organogenesis were not available in the old text (Astrology). So, it has been developed on this pattern for therapeutic purposes.

Another major similarity falls on spinal segments. The 360° sky was divided into major 12 parts as rashies, but for minute characterization of stars of the whole sky was again divided into 27 parts known as constellations. In the human body also, there are 27 important spinal segments. There also similarities are found, i.e., 1st spinal segment found attached to the same organs as 1st constellation indicates, so on, so forth approximately. Thus organogenesis was completed with the partial help of old science (Astrology). During therapy, this pattern was found correct. Genetic discussions will continue, in the meantime, details of the important papers enclosed are given below:

1. Ethical committee-according to ICMR guidelines, was formed comprising of a Retired Judge, a Physician, a H.C. Advocate, a Mukhia of village panchayat. It was constituted for legal monitering of human patients during therapy (cf. chapter II.)

2. Call Letter of AIIMS (Dated: 20/10/93), New Delhi after many discussions by post.
3. Letter of science council, Japan, President placed the matter in its Liaison committee. Chairman Yukio Hiramoto, and chairman Ikuo Tekeuchi of zoology and botany respectively reviewed and commented.
4. In the Letter of Liaison committee Science Council Japan 9/1/1989, they declared the subject, new to science and wished all success in the mission. (cf. chapter I)
5. Director General, ICMR, New Delhi asked to submit a project on the topic. Astrogenetics (a sample study before and after Astrogenetic regulations), in collaboration with genetics department, AIIMS (8/2/1996) (cf. chapt. I)
6. Letter of Vice Chancellor L.N.M.U., 1/11/1993 for submitting the list of infrastructural requirements for Astrogenetic research. - (cf. chapt. II)
7. Scientific paper published in Environment Ecology 10(1): 117-123, 1992. ISSN 0970-0420. Effects of some naturally occur chemical compound on human diseases (Allergy and others). (cf. chapt. II)
8. N.L.O., Bulletin News, page no. 158 (cf. chapt. II)
9. AIR Broadcast 20/8/1991 "Ratna chikitsa dwara rogon ka nidan aevam nivaran". Recorded on 23/8/1991 at 2pm and broadcasted at 8pm. (cf. chapt. II)
10. AIR 21/11/1991 (Letter), 29/11/1991 at 2pm (Recording) and 30/11/1991 at 6.45 am (Broadcast) "Cosmic kirno ka sharir par prabhav". - AIR (cf. chapt. II)
11. Hindustan (Newspaper) Publication 3/1/1995 cf. chapt. II
12. KADAMBINI National magazine "Ratna chikitsa ki ek anokhi padhati-Astrogenetics" December 1994 (cf. chapt. II)
13. Times of India, Darbhanga 15/1/1995 "Miracle cure of Liver Cirrhosis" (cf. chapt. II)
14. Gaandeev, U.P. 21/1/1995 "Liver sutran rogi ka chamatkarik Ratna upchar se labh" (cf. chapt. II)

15. Times of India 15/1/1995 "Stone researchers claim to cure terminal ailments" (U.P. edition cf. chapt. II)
16. Hindustan, Patna 15/4/1995 "Astrogenetic namak nayi chikitsapadhatti ka izaad." (cf. chapt. II)
17. The Indian Nation, Patna 14/7/1995 "Astrogenetic cure for terminal patients." (cf. chapt. II)
18. Aaryavart, Patna 13/6/1996 "Rogon ki chikitsa me ratno ka chamatkarik prabhav." (cf. chapt. II)
19. Dainik Jagaran 18/10/2000 "Ratno se lailaj rogon ka upchar sambhav." (cf. chapt. II)
20. Aaj, Darbhanga "Ratna patharo ka manav chikitsa per sarthak parinam- Dr. Jha." (cf. chapt. II)
21. IIT, Bombay 12/11/1997 Confirmation Letter for Genetic analysis from 21/11/1997 to 7/11/1997 free for mutual interest. (cf. chapt. I)

Any member of the society never contributed any news item to newspapers or magazines. They were directly obtained through patients or available literature of the society but scientific discussions and papers were the sole contributions of the Director. Rarely reporters organised news-conference with patients in the society may or may not be in presence of the Director. AURA which was our point of discussion, revealed, attraction repulsion force which however, at a very large scale apparent in the solar system where planets behave similarly. They are balanced through the similar type of attraction (gravitational) force but only between nonliving. This is the beauty of a natural phenomenon that more or less same set of mechanisms take place both in small and large objects. It is also miraculous that a big effect is produced through a minor event. This is the rule of nature and beauty at the same time. Regarding Aura, we discussed at length, but the gravitational force is yet to be conceived on the platform of astrogenetics. The gravitational force develops after the effect of cosmic rays on tiny molecules. The inner electron jumps in regular fashion, producing very small magnetic field.

Such reactions taking place in huge surface of the earth. Might be the reason of so powerful gravitational force developed on the planet Earth. It is evident by the fact that planets may differ in other aspects but gravitational force is created in all planets with only two similar things under interaction, i.e., cosmic rays and chemical/mineral surface of the planet, these two are only common, (matter and energy). So gravitational force appears as the outcome of the interaction of the two everywhere in the cosmos, may be low or high. Usually there is acrimonious debate on the presence or absence of the Almighty. This is very easily settled on the platform of astrogenetics. The effect of cosmic rays and their interactions behave supernaturally through aura, gravitational force although scientific. The God is present or absent none can say emphatically because one galaxy contains many solar systems, and more than a lakh galaxies are in the record, which is still expanding. So it is just possible that God may be anywhere because noone has travelled the whole universe or vice versa also be true. Nobody can claim it as positive or negative. Even if He is present in the cosmos, He would have been influenced by cosmic rays (control system). He in that way might have obeyed the effects, that our mythology also confirms. So, this mechanism annals the dispute and puts healthy conclusion on one's mind in relation to religion, without putting any stigma on science. Consequently this cosmic rays mechanism balances both science and religion, unveiling greatest truth through genetics and aura mechanism.

Solar Spectrum

Astrogenetic observation regarding the solar spectrum- The solar spectrum as in the light of the modern science consists of the following:

VIBGYOR

V - violet

I - indigo

B - blue

G - green

Y - yellow

O - orange

R - red

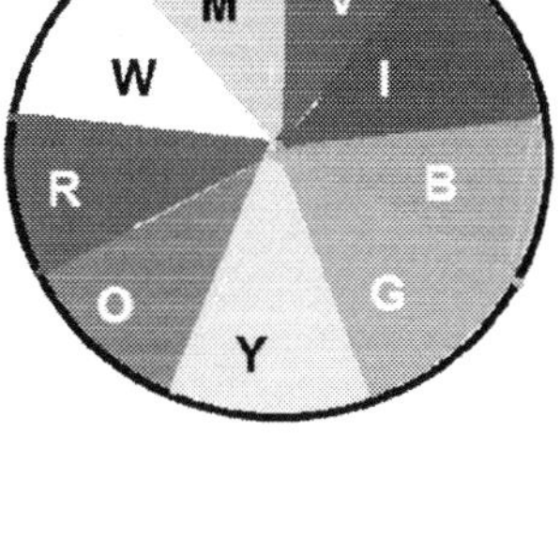

According to ASTROGENETICS

VIBGYOR + MW

White **(W)** is the base colour. The source is very remote coming from stars of Venus characteristics.

(M) is Mercury whose colour contributed by the moon reflection. So, a spectrum with these colours gives more accurate effect of ray. It is an astrogenetic belief. Stars are the real main source of rays. So, astrology is the science of stars. Since, during characterization, traits of groups of stars assigned as planets. So, astrologers were confused, they started telling that it is a science of planets. But the truth is that planets are simple chemical/mineral surfaces placed in the way but rays coming from stars, interacting and changing their characteristics for further interactions with living and nonliving systems. The Sun is also a star. So rays contributing bodies are stars and galactic nebulae. Planets never emit rays. So, they are never rays-contributing, but they are of different constitutions giving good or bad effects to the rays. It has been experimentally proved that rays are imitated, when passed through different solids and liquid media. The change of an angle due to refraction also causes a lot different effects. This is causing alterations, i.e., change of characteristics.

Characterization of rays has been done in the following manner:

Rays of 360°, divided in 12 parts of sky

1st and 8th parts bear characteristics of planet Mars	- from both studies Astrology and Astrogenetics.
2nd and 7th bear characteristics of planet Venus	- from both the studies
3rd and 6th bear characteristics of planet Mercury	- from both the studies
4th that of Moon	- also similar in both
5th that of Sun	- similar in both

9th and 12th that of the planet
Jupiter - similar in both

10th and 11th that of the planet
Saturn - similar in both

The characteristics of stars of a particular part of the sky have been assigned characters of planets. That becomes simple to cover up characterization of large interactions (more than 85 lakhs). The most striking and important fact is about the sunrays; it does not consist of six/eight other rays shown by the spectrum rather it is a single red part of the spectrum. All other rays have been contributed by different parts of the sky. However rays coming from different parts of the sky interact on the surface of sun, producing sunrays with seven/nine coloured spectrum. Nine is anastrogenetic correction. According to an astrogenetic conviction that Sun contributes only the red part of rays in the spectrum and all others are contributed by stars of different parts of the sky.

Cosmic Rays

Something has already been told. Yet, some more details are required since it is the main force playing vital role with matter of living systems. Sometimes it is confused as almighty force (or fantastic rays) but definitely it is a scientific entity atleast now when studied in detail through Astrophysics. According to an American reporter, Victor Hess, in 1912 with ionization meter observed at 16000, feet the force of these energy particles. Then he reached to the conclusion that the sources of these powerful rays are sun, stars and gallectic nebulae from around 360° of earth. Energy is variable. It is 1 to 1 lakh Bevistron. It has been found on all planets. Scientifically it is a proton. This is present everywhere in the universe. That's why it may be assumed as a fundamental particle which might have played role in evolution. This is the only energy particle present in universe so manifesting various types of energy found on earth, e.g., electromagnetic forces, etc, we perceive. This energy is developing gravitational force and environment of the earth. Victor Hess observed it on such height through a balloon flight. This does not necessarily mean that the sun is the sole source of cosmic rays. If it would have come only from Sun, the intensity should vary markedly with the position of the sun in the sky. This variation is negligible, less than half per cent during 24 hrs. The earth's magnetic field makes cosmic ray particles of moderate energy, it thus produces a more or less random distribution in their direction of arrival. But the magnetic field has no appreciable effect on particles of higher energies, and yet they too appear to arrive with almost an equal intensity from all parts of the sky. So from all relevant discussions it is proved that the Sun is not a sole source of cosmic rays (Scientific American). As also told on the basis of A.G. experiments, the only red part of the spectrum is contributed by the Sun. All others contributed by the stars of different part of the sky. Teller, also postulated that kinetic energy is exchanged for magnetic energy, i.e., gravitational force of the earth, as told earlier with a slight change in the mechanism. A Swedish Astrophysicist, Hannes Alfven also proposed a mechanism to prove Teller. These views of modern science tally with the astrogenetic concept, producing an equilibrium between science and religion with certain scrutinies on both the sides.

From foregoing account, it may be derived that cosmic rays, besides control mechanism, induces environments and rarely cause mutations. This is now high time to ponder over that. We are all surrounded by cosmic rays, but modern science ignores it. Even life science does not consider its effect in any disease or on life forms. Is it not a hypocracy? Due to the fact many diseases remain undiagnosed and untreated. Natural calamities are not always reported correctly. The main shortcoming evidently reflects in the silence which appears escaping definite causes at many instances. Usually doctors do not categorically mark causes of the most of the diseases. Sometimes, their arguments become fallacious. For example, according to them lever cirrhosis is caused due to heavy intoxication but numerous cases appear without taking any bit of intoxication. That's why there are many confusions in their claim regarding causes of the diseases. So, it is relevant to comment that still life science (Medical science) is on an intermediate path still away from real causes of the human ailments. That's why many diseases are yet to be cured.

Cosmic rays are angle, power, and interactions specific. When an infant's head first of all oozes out of the mother's womb, that is the first appearance of the baby in the universe. Instantly the head of the baby comes under the control of a particular part of the sky that is considered thefirst part of the individual, that may be any part of the sky (1 to 12). It is calculated at that particular time and place. Time refers to the planetary positions in the sky at a particular moment. Local time noted at the moment of the head oozing is converted into standard time, and with the help of latitude-longitude of the place, planetary position is computed to prepare one's cosmograph.

Cosmograph so prepared, scientifically requires primary factors, i.e, time of birth, parents and place of birth. These factors make it different from millions. Is it not fact? The position of the planets at that particular moment of birth, switches on the clock of his life. As the path and movement of planets are fixed, further calculations may be derived and calculated for that infant of certain genetic characteristics (parent).

Angle

In the cosmograph computed at particular time and place, the aspect of the first house when determined under control of a particular piece of the sky, e.g., 6th then controlling planet becomes clear from old science Astrology. The presence of planets in that part of the sky is already worked out from observatory. So, subsequent parts after the first may easily be computed and then planets are placed in each house at the time of birth. If starting from the first house of the individual, i.e., 6th (for example, cf. Astrology and observatory - planet chart, i.e., Patra)

- 7th becoming 2nd,
- 8th becoming 3rd,
- 9th becoming 4th,
- 10th becoming 5th,
- 11th becoming 6th,
- 12th becoming 7th,
- 1st becoming 8th,

- 2nd becoming 9th,
- 3rd becoming 10th,
- 4th becoming 11th,
- 5th becoming 12th.

Owner planets are the owners of these angles and houses. Among them few angles and houses are deleterious as such the planet of that house is considered bad. It is established that 2nd, 6th, 7th, 8th are bad angles and positions. Out of them, 6th, and 8th are very bad. The owner of 8th regulates loan and death 6th controls disease and enemy. While 1st, 4th, 5th, 9th, 10th, 11th angles and owners are considered good, 3rd owner and angle are partly bad.

Power

1st, 4th, 7th and 10th houses are powerful, may be 80, whereas power of 5th and 9th may be 100. 3rd, 6th, 12th are weak (may be 70). Houses, owner and placed planets are considered weak. There is a point of confusion, so a bit concentration is required. The normal sky has been divided and assigned as the first part, to Mars, 2nd, to Venus, 3rd, to Mercury, 4th to Moon, 5th to Sun, 6th to Mercury, 7th to Venus, 8th to Mars, 9th to Jupiter, 10th to Saturn, 11th to also Saturn, and 12th to Jupiter. This is for all individuals. But as one individual takes birth and his first house was computed to be the 6th part of the sky, then, as told above, that will be first part and so on and so forth. Here one should be adjusted with two cycles I, normal, for all individuals and 2nd personal of the individual. This causes individual difference.

Interaction

Rays coming from a particular part of the sky may interact in the way with 1 or more planets. Those planets appear to be placed in that house. The characteristics of planets interact. In this way, it shows different interactions and result differently. The result of interactions has already been worked out. For example, rays of the 2nd part interact with planets placed Saturn and Mars, then it is the Venus, Saturn, and Mars interaction. If

4th house is occupied by the Sun and Saturn then it is Moon-Sun-Sat-interactions, and so on.

Reflection

Rays of the parts of the sky falling on the planets may reflect. The angle of reflection is fixed. As that of Mars, 3rd, 4th, 5th, 7th, 8th houses:-

Venus	3, 5, 7th
Mercury	3, 5, 7th
Moon	3, 5, 7th
Sun	3, 5, 7th
Jupiter	3, 5, 7, 9th
Saturn	3, 5, 7, 10th

They also form interactions. If in the first house of cosmograph, and 7 number is marked and Mars is present in the house then it is interpreted as Venus, i.e., owner of 1st house, and Mars is placed in first house, so Venus-Mars interaction is present in the individual. Likewise, interpretation goes. One thing is also remarkable that Venus-Mars interaction is specific. The effect of such interactions has already been worked out to interpret. Thus, we have discussed basics of interactions, reflections, interaction due to the presence of more than one planet in one house. As there are more than 85 lakhs interacted rays by different interactions, among themselves and with planets, so they are causes. Effects are produced after interactions with matters.

Gradually, it appears that different important interactions of rays affect the whole genome of the individual present in the whole cosmography of the individual and playing phenomenon of genetics like dominant, recessive, coupling repulsion, crossovers and genetic transmission of character,s etc. to establish characters of the individual. This phenomenon will be displayed when pedigree Charts of 2-3 generations will be discussed with cosmographers. Then genetics will be proved by the mediation of cosmic rays. Direct experiments of cosmic rays are difficult, since not even pinhead space be created cosmic rays free on this earth. So, experiments are conducted by increasing the rays and also on effect to cause analysis pattern. It is well-known that the sources of cosmic rays are very remote. Direct experiments are very few. In this connection, I would like to quote Puspa Shrivastava's experiments on few samples of algae. She exposed the sample in cosmic raditions, i.e., at higher altitudes in organised balloon flights. The exposed samples of different durations revealed, basically three types of effects, which are as following:

1. Beneficial
2. Devastating
3. Mutational

So, now people will think in this direction, and in near future, life scientists will rush with experiments on cosmic rays. This is definitely a good news. The modern science was paying no heed rather deaf ear which was a hypocrisy of modern science, thus deplorable. The specific chemical mineral composition of the planet Earth after interactions with cosmic rays, initiated the process like chemical evolution. Thereafter in due course, the development of basis of life, i.e., hydrogen and oxygen formed which after taking energy from cosmic rays might have developed first water. This water could have undergone reactions with chemical (mineral) surfaces of the planet under the effects of temperature ammonia lightening etc. to produce complex chemical compound like Protoplasm which is the physical basis of life.

Then evolution of life started, but at every step, the role of cosmic rays was vital either through energy support or through manifestations. Effects of interactions of rays and planets have already been dealt in different books of astrology. They are actually the effects of the interactions of genetic materials and cosmic rays (matter-energy). There, it is not depicted likewise; simply because of religious concept. However, modus operandi is scientific and results are same. It is well within doubt that genome study as going on these days may culminate in similar fashion, but easier and in non-expensive way by astrogenetic means. Trait analysis is going on in these two (genetics and astrogenetics) is more or less in same way, but with sophisticated tools in life science. So, convergence of these two may pave ways for more minute, results in very simple way. Powerful interactions will definitely result in more or less same through both processes; (Genetic and Astrogenetic) such as genetic traits of these powerful interactions:

Sun-Mercury, Sat-Jup, Rahu-Jup, Ven-Mars, Moon-Jup in 4th normal part of the sky, i.e., moon, provided they are present in houses with 100 or 80 powers result in acute bad/good effects are also bound to manifest both the systems. Peculiar confluences are also found to reflect.

Apparently interactions are all due to the following:

1. Part of sky (Rashi) with planet (as 12 x Sun)
2. Planet with planet (Sun x Mercury)
3. No. of house of the individual with planet. (individual 6th x Mars)

These appear to be genes under action of cosmic rays as because whole cosmograph represents complete genome. Planet to planet interactions necessarily require cosmic rays, because it is taking place in the part of sky (Rashi) where one or more planets transit. It is always to be remembered that planets are not rays contributing, rather playing a passive role by only reflecting rays, of course manipulate through chemical actions. The effects of these interactions are well elaborated in books, appear more or less correct forming parts of old testaments of experiments. Astrogenetic rectifications are few, mostly confined to therapeutics. They will be mentioned during a particular discussion. Characterization of each 30° of the individual, i.e., houses are computed from birth time and place of the individual.

- **1st house** of the individual may be any between 1 to 12, say 7th, or 11th any on calculation 1st house-denotes personality of the individual, illnesses, wellness, nature, longevity, character, mind. It also refers to organs which will be discussed in organogenesis.
- **2nd house** depicts wealth, materials, relatives, beauty, pleasure, truthful, hard wealth, love, singing. Its owner plays evil role giving evil effects.

- **3rd house** depicts victory, brothers, social, servants, patience, ornaments, exercise, etc. Astrogenetically these genes are regulated by the stars of third part of the sky, of the individual not general. However, when both are considered, more accurate results are received, i.e., individual's 3rd part and general 3rd part mercury, both taken into consideration. It also refers to organs which will be explained in orgenogenesis in health, wellness section of every part.
- **4th house** is powerful 80°, depicts pleasures, vehicles, mother, garden, friends, peace, parents pleasure, service, transfer, service to humanity or devoted for others, etc. These genes are regulated through individual and general 4th part of the sky (Moon).
- **5th house** regulate genes of good or bad natures of following traits, e.g., issues, study, wisdom, management, speaking words, legal management, patriotic, supernatural devotion, fame, sudden wealth gain, service termination, mothers blessing and pleasure, etc. This is also a powerful house 100°.
- **6th house** emancipates disease, enemy, envious attitude, maternal uncle, worry, suspicion, landlord ship, etc. organs will be described in organogenesis.
- **7th house** regulating working capacity, opposite sex, marriage, business, sexual potency, etc. this is also powerful house 80°. It interpreted both from general 7th and individuals 7th.
- **8th house** Longevity, loan, life and death, travel, voyages, etc. situation of planet Mars here is very adverse for opposite sex; accident.
- **9th house** religion, fate, vices and virtues, father shelter; concentration, pilgrimage, living outside home, mental work, sacrifice, character, etc.
- **10th house** regulates the genes promotion, politics, business, obedience, respect, father, omission and commissions, etc. It is also powerful house with power 80°.

- **11th house** is the star of this part of the sky that regulates income, benefit, horse, elephant, precious stones, vehicles, auspicious fiction, luxury, etc.
- **12th house** regulates traits like expenditures, law court, foreign, gift, donations, loss, punishment, bad habits, etc. In every case, interpretation is drawn on the basis of study of general and individuals no. of houses.

Positive, negative and magnitude of the effects may be judged on the basis of malefic or beneficial effects, e.g., it depends upon gross effect. If characteristics of stars are supported by the same characteristics of the planet present there and no planet present; or adverse planet present make the effect powerful to adverse. So, after all, interpretation in right direction is required for full correct results. It is apparent that gross effects of stars of same characteristics i.e. in particular part of the sky act as weapons to affect part of genetic materials (genes) of the individual to yield results switching on the activity. The result depends upon the quality and nature of genetic materials an individual possesses. The variable results among individuals are evidences of correct interactions and correct results. Thus it approves the phenomenon. The genetic characteristics are evident in real cosmographs plotted as figures G (1-8).

Astrogenetics and Precious Stones

It is a pertinent question that man of every walk of life uses precious stones since days Immemorial. Kings and emperors were fans. The story of *Kohinoor* is well known to all. Emperors were mostly using precious stones for the wealth purpose. *Ayurvedacharyas* use them for medicinal purpose, but they were damaging them without knowing their dynamic exogenous effects. *Ayurvedacharyas* were burning them to make Ashes (*Bhasmas*) for getting tremendous strength in convalescence, aphrodisiacs like uses, and inducing killing effects. It is still prevalent, but such uses reduced due to high cost. How unfortunate this is, that for one patient few precious stones are burnt. This becomes not available for others. It is recalled that these stones have special trait to be used exogenously on body, i.e., effects be obtained without damaging through the action of rays on them so that others may use. The matter illustrated was not out of place to be mentioned; however focus on the mechanism of their production in Nature itself speaks many things. We already know that prior to organic evolution, there was chemical evolution.

The process is still going on. The cosmic rays play a role of "energy" support during evolution and Earth's chemical as "matters". During the course of evolution, a particular ray constantly fell (interacted) on the piece of earth's chemical as whatever depth it may be for thousands of years converting it into chemical compound which certainly develops a trait to capture those rays from environment even on the depleted surface. They may change bad angles of rays, usually remain in mixed condition so compounds selectivity makes rays pure. These pure rays reach to the organs or genes of depleted energy status of the individuals. Duration of action of ray on the chemical surface differentiates it into precious or semiprecious nature of stone (RAD) ray altering device.

Cosmograph

We will discuss it under two heads: I nature II development and computation.

Point I:

Nature is discussed here simply because its characterization has already been started. So, its relevance both with topic and nomenclature with genetic behaviour must be illustrated to justify nomenclature of the whole episode as astrogenetics be brought to light. So far its development (scientific) and computations are concerned, it is already with statistics in chapters ahead.

The whole cosmograph genetically resembles human genome both in traits and behaviour. It governs characters (traits) and displays genetic phenomenon well illustrated through some plates of real persons and families. As this whole genome comes under cosmic rays since inception of a baby on this planet Earth, the genetic drama starts thus characters evolve abiding the rules of genetics as well. The complete act of an individual is completed in cosmograph by the mediation of both genome and cosmic influx, Hence, the nomenclature Astrogenetics was assigned and resultant genetic phenomenon were plotted in cases studied. It is also claimed that in every study of family cosmographs, genetic behaviours will be witnessed either very specific or it may be simple through real graphics of rays (pure/ interacted) displayed in cosmograph.

Factorial Presentation of Birth-chart Cosmograph

For Scientific and genetic explanations, supporting astrogenetics is composed on the basis of astrological (indian) tenets as in birth chart.

Horoscope Computation and Setting A/C to Indian Astrological Tenets

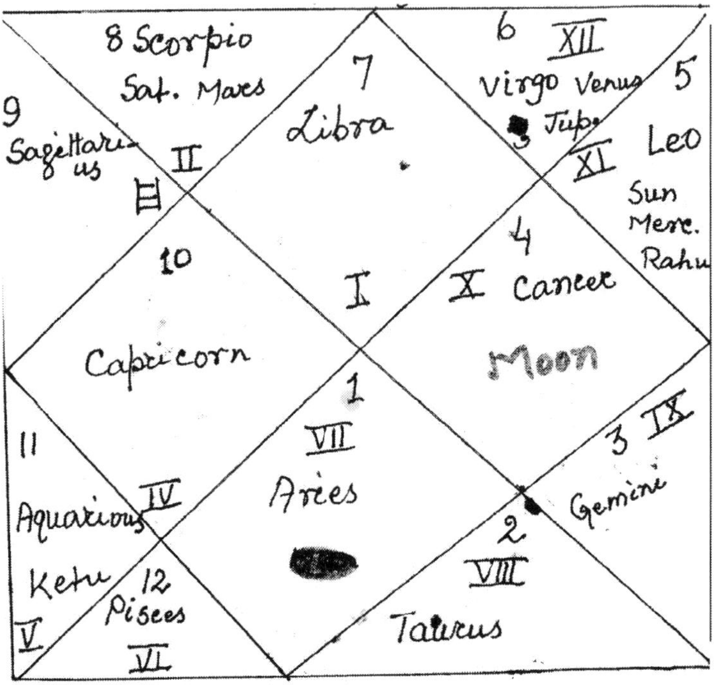

A. No. of Houses One characteristics (Starting from Lagna)	B. Serial & No of Rashies :- Another characteristics	C. Planets :- Third characteristics
I	1. Aries	1. Mars (1,8)
II	2. Taurus	2. Venus (2,7)
III	3. Gemini	3. Mercury (3,6)
IV	4. Cancer	4. Moon (4)
V	5. Leo	5. Sun (5)
VI	6. Virgo	6. Jupiter (9,12)
VII	7. Libra	7. Saturn (10,11)
VIII	8. Scorpio	
IX	9. Sagittarious	
X	10. Capricorn	
XI	11. Aquarious	
XII	12. Pisces	

I. Minute Interactions and Characters:

1679616x27x27

= 1,22,44,40064

II. Broadly Interaction Characters:

(9x9x12x12x27x27)

= 8503056

III. Interactions:

12x12x12x12x9x9

A X B X C = 1679616

Three examples have been presented, and elaborated as live testimony to substantiate that Cosmic rays mediate genetic display and inheritance of factors/genes/characters in human being. The source/route of factors/characters also visualize either from mother or father (well depicted through cosmograph). Though there are numbers of examples showing these bewildering facts have been compiled. Some of them will be cited again after a detail description of cosmograph. What is this and how is it computed?

Origin and Statistics

Now origin and statistics of cosmograph will be illustrated. How is this new nomenclature adopted and what is statistical back ground?

Cosmograph is a factor based horoscope, for displaying, Interactions and inheritance of characters very easily and scientifically to even laymen ignorant of astrological terms (for scientific analysis). Here astrological tenets are never manipulated in pure scientific/genetic pattern. So, one solar science prevalent some 5 thousand years ago among 18 sciences tried for characterization of cosmic weapons in relation to man. It further explored for the knowledge of mechanism of the earth, i.e., mechanism of TRIO. Cosmic weapons (Cosmic rays),+living system (man)+Non-living objects. Very clear relationship obtained through aura science, already explained previously. That also formed the basis of experiments, which ultimately proclaimed astrogenetics, after thousands of cases studied scientifically under Astrological principles and scientific tools and techniques. In the midst, a bewildering fact became crystal clear that cosmic rays are the causes of all human diseases and genetic phenomenon. Consequently, astrology which was known as occult science, now used for perfect scientific analysis after getting empirical results. Cosmic interactions thus may be called as weapons for handling intra and inter relationship of living and non-living systems. As such, this Astrological science is explored scientifically to characterize cosmic rays. The horoscope depicts a picture of the individual taking birth at particular place (altitude and latitude) with all planets falling in *rashies* and constellations at that moment of time when one's head comes out first in the touch of the environment. So, this horoscope is simply a picture of individual in relation to that moment of time. Time and place play most important role on genetics, i.e., a child of a particular parent i.e. of particular genetics undergoes developments, whereas infant of other parent in the same moment of time and place develop quite different. So, factorial and scientific casting of horoscope is called cosmograph.

Cosmograph as displayed appear to have genome control in the whole surface as the baby has developed and attained genome contribution of parents well within womb. The next major role of only cosmic rays takes place when head of the infant came out of the womb. Pertaining to the fact underlined, only two main actors participated for the whole journey of life of the baby. They are notably genome of parent characteristics and cosmic status switching at the particular moment and revolving there after giving changing characteristics in the scientific pattern of time. This is remarkable that parental contribution in particular time and place (altitude-latitude) makes baby different from all others. This is individual difference (except in identical twins). Thus specific features of time and place for new born are scientifically understood in brief. However, the relevance of genome waits. The whole cosmograph seems acting with the aid of cosmic energy (of planet characteristics and with planet itself) no doubt but in that play main body acquired from parents. Their behaviours are well depicted in figures (41-8). Cosmography contains two clues, one as human being (general) and other as particular individual. As human being, it is labelled as 1-12; according to the parts of sky contributing cosmic influx (already described) MA - Ve - Me - Mo - SU - Me - Ve - MA - J - SA - SA - J. Suppose any individual (new borne) according to time and place took birth and caculated as Lagna Me IIIrd. This Me, becomes his personal - I- (L) in addition to his general Ist is I rashi (Aries). Thus his Characters of 1st part of genome or genes will develop taking both the traits of MA (general) - ME (individual). So on so forth, for other parts of genome, ie., it includes 2nd to 12th parts. With this vigilant eyes when viewed, several genetic behaviours (phenomenon) become apparent in cosmograph (cf. figures G1 - 8). On the basis, it was proclaimed as astrogenetics and boldly asserted before high scientific institutes. Then question rises why studies on human genome is not conducted through this simple study instead of huge expenditures. Exploration with some scientific tools through astrogenetic route may pave tremendous success in no time and expenditures.

The following illustration and graphs will appear fantastically true when whole cosmograph with interaction is considered as genome (already discussed).

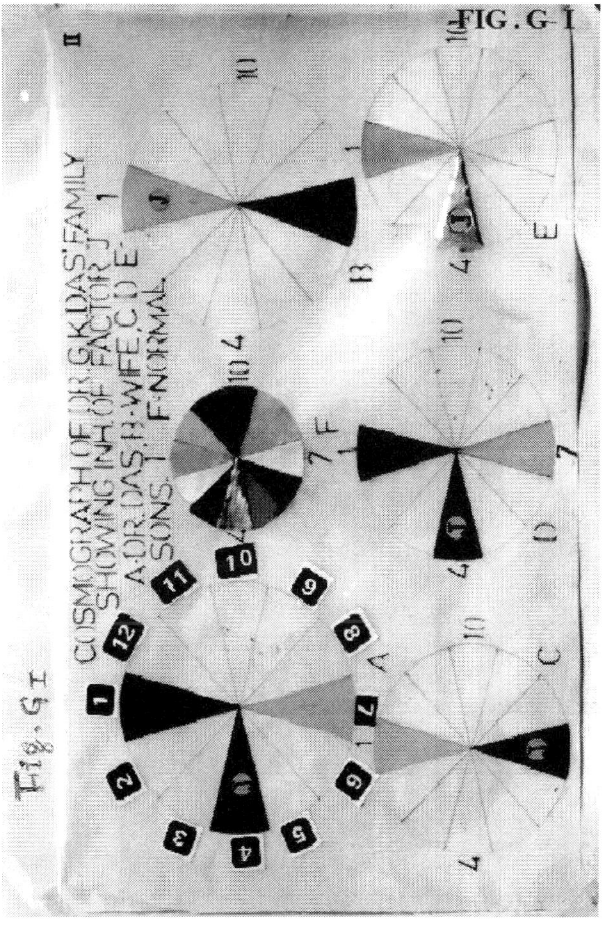

Fig. G - I

This figure depicts Genetic inheritance with social reference of a particular dominant trait. This figure displays. The cosmograph already plotted pretty long before with all real and scientific statistics of parents and three sons of a Doctors family (Dr. G. K. Das). The factor J is present as dominant character in both parents in 90d powerful houses. Further the factor (J) apparently inherited as dominant character in 90d powerful houses 7th & 4th in all the sons. Consequently all the sons were well placed in life as IAS, professors, doctor, etc.

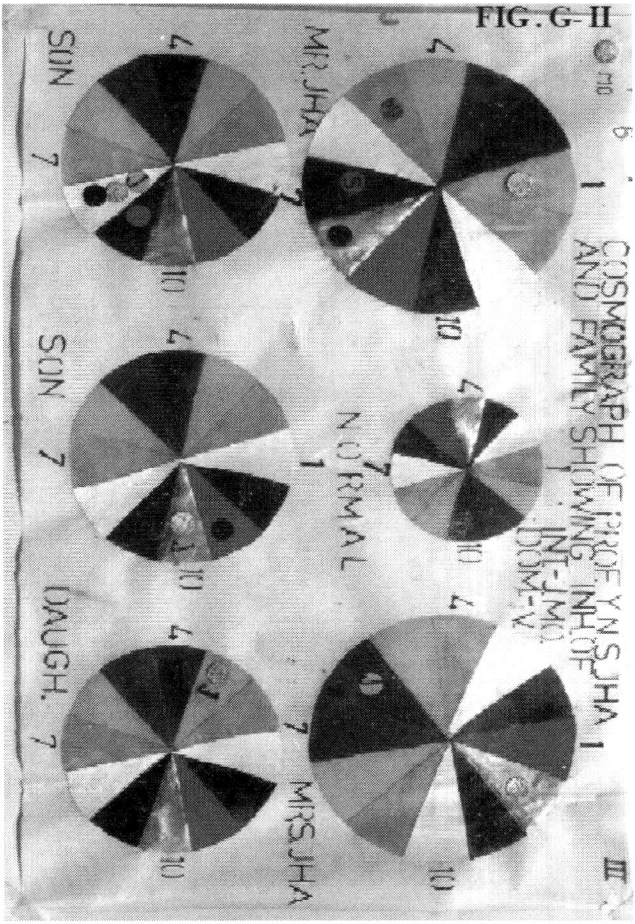

Fig. G - II

The second plate substantiating the same fact of inheritance of some other factor in the family of a professor with wife, two sons and a daughter. In father factors J – Mo is present in

1st 90d house. In wife this shows at right angle interaction in 6th & 12th 70d. Any how this confluence is present in both the parents. So, this confluence is inherited in all childred. In 1st son, it is in 8th (J- MO). In 2nd son, it became dominant like father in 100d house and in daughter this confluence also appeared but in 70d power. Likewise, future development of children appeared. Father and younger son developed in the same developmental strata but daughter and 1st son (weak in both) not progressed so well. This study elucidates over both the phenomenon, genetics as well as development and differentiation of genes, studied in such a simple way.

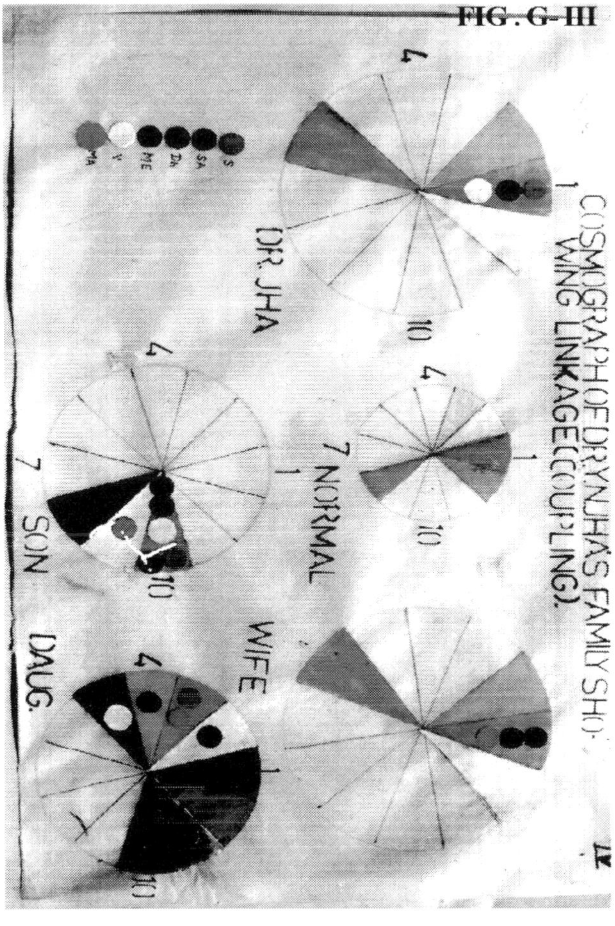

Fig. G - III

Genetic phenomenon coupling is well displayed in this figure. In male parent three factors are (90d) prominent (S. SA. MO.), and in the female parent other three factors are prominent (90d). Consiquently, there in coupling of 5 such factors from both the parents when one is showing repulsion in son. The factors were powerful in both the parents (90d). So, they were inherited powerful in the son (90d). But in the case of daughter those factors spread (Segregated) near by. So, some became weak and some powerful.

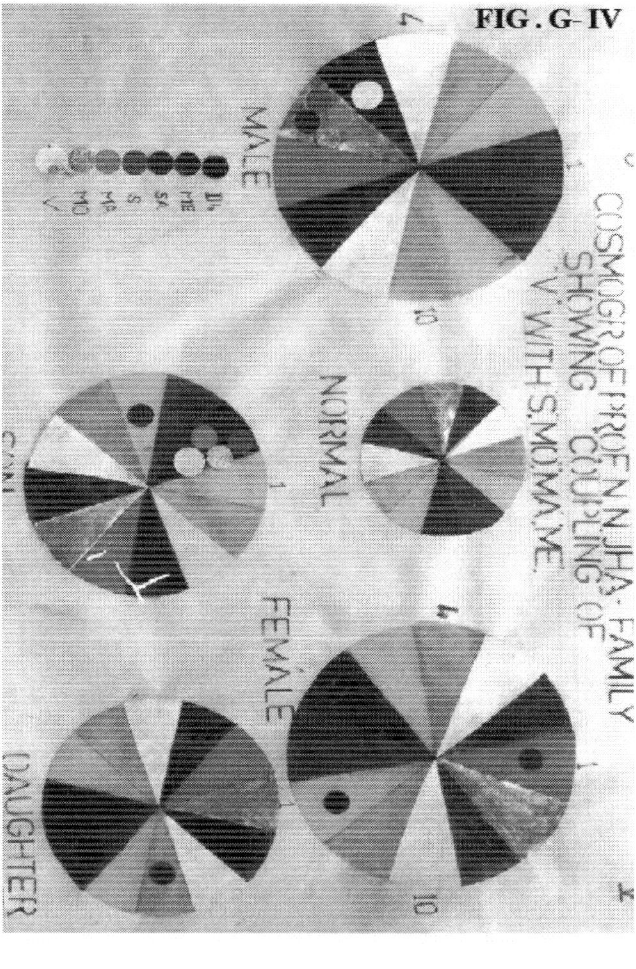

Fig. G - IV

Again coupling phenomenon is well displayed in the members of another family where some traits of the male parent and some traits of female parents assembled together in one house of the son.

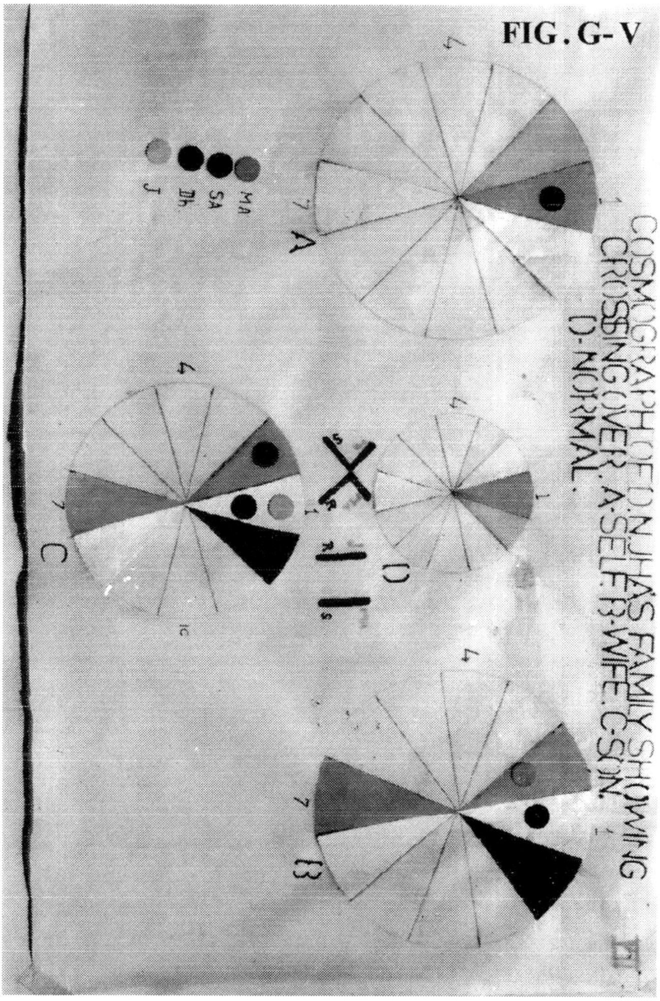

Fig. G - V

Marvellous phenomenon of cross-over was witnessed in this plate. This true plate was prepared with cosmographs of parents and a son. Eventually cosmographs were plotted at variable times. In the male parent, characters (trait) J (yellow) - S (black) are situated side by side in 80d and 90d powered houses where as in female parent trait M (orange) and R (black) are situated side by side, M with double power in 80° house but R in 90° house. Now in son, the traits appear like J - R and M - S, showing clear cut case of crossing over.

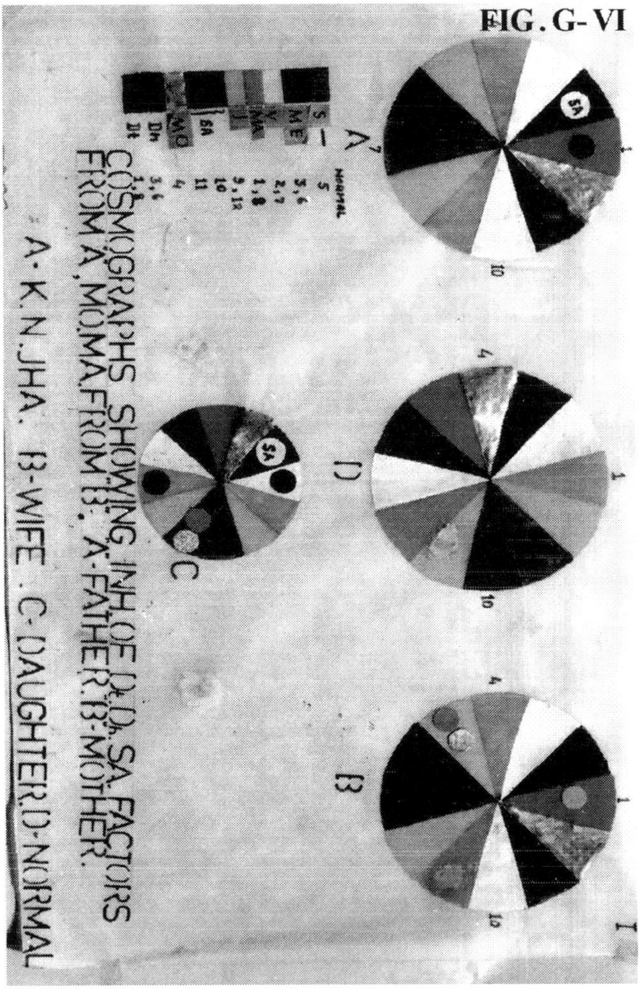

Fig. G - VI

Inheritance of few factors are well depicted in this family cosmographs, JS x MR JR x MS. In the cosmograph of son, it is clear that Dt, Dh and SA factors are contributed by father, MO and MA factors received from mother. SA factor was received in son in the same position with same power from father while MO - MA factors received from mother in another but with same power house. It is also highly remarkable that both parents first house is regulated by S factor.

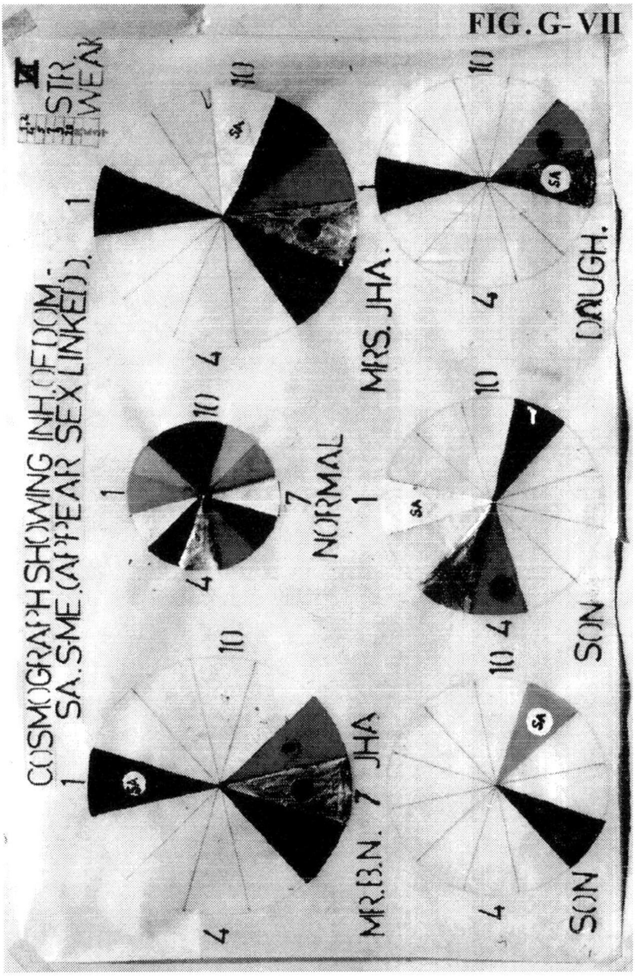

Fig. G - VII

The first house of both parents are regulated by S A factor which did not regulate 1st house of both the sons in the same way. However, the daughter's Ist house was regulated in the same way as in mother. So, this inheritance of S A in daughter appears sex Linked. S - M E interaction developed in both parents but it is carried over only in one son (2nd) at 90° from father. It is highly remarkable that all genes/genetic material/ houses of both parents are under the influence of same planet characteristics might be the reason of becoming wife- husband.

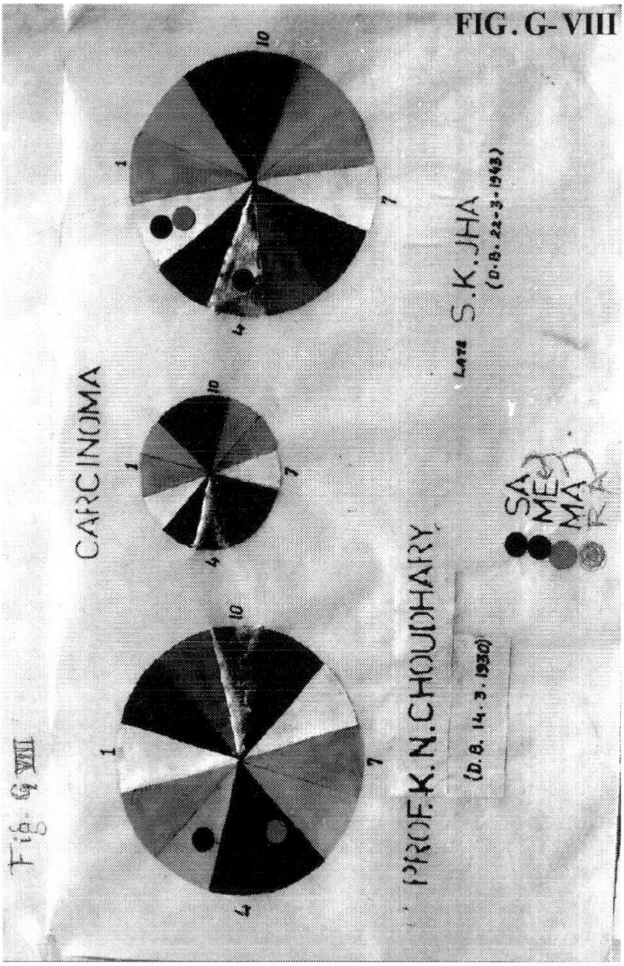

Fig. G -VIII

Highest deleterious interactions so far studied in the interaction causing carcinoma. This plate exhibits two cases in the first case SA - MA is present in 3rd. This interaction when comes under the influence of powerful bad RA, the patient undergoes malignancy. RA may be permanent in the interaction or transitory. So, the place/house where interaction of all malefic planets SA - MA - RA converge/interact and carcinoma develops. The organ involved is in accordance with the house of interaction.

In second case, SA - MA interaction is apparent in V (house). RA will be transitory. V and 2nd house denote organ involved. In the 1st case, organs will act according to organo genesis of 3rd house and MA.

Thus, up till now it has been observed that carcinoma is a result of interaction of 4 planets. 3 active - SA - MA - RA and one induced or passive, i.e., the house or factor denoting organs. This usually remains place of interaction. House means part of genome. Involved planets as and when develop malefic property trouble starts. So long, they are not malefic and full interaction have not been completed, they are known as pro onco gene but as soon as interactions are completed it is converted into onco gene responsible for cancer.

SA - MA - RA - + organ factor

On the basis of some live evidences conclusion may be drawn that cosmic rays may be causes of Genetic phenomenon. In daily life some very remarkable results appear in the cosmograph of mother and Daughter. One daughter is very much similar to her mother phenotypically, their cosmographs also strikingly. Similar but different with the cosmograph of another daughter which is not at all similar with mother in appearance also. These are cases encountered every now and then but we ignore.

As discussed earlier:

The rays Altering device, which developed through above process are called precious stones (*ratna*), i.e., through the constant exposure of a particular part of the earth mineral or chemical under particular ray for thousands of years brings chemical evolution and converted as precious stone (RAD). NATURE is sometimes very divergent and in the way it manoeuvres many required things. So, some other parts of the earth's chemical and mineral also become exposed to particular rays for lesser, periods. They also develop rays trapping and altering property to very lesser extent and depth i.e. not acting on third layer (i.e. gene level) are known as semi-precious (upratnas). By birth some individual's become deficient in some rays and getting evil effects of some other singly or in mixed condition. Individual takes birth at particular time and place. So, he does not get some rays properly, some may be due to bad angles and some also due to mixed condition. The RADs make them available to the site of requirement in good angle and power for gene action. Now it depends upon stone prescriber to select precious or semi-precious depending upon the requirements.

3rd, 6th, 7th, 8th and 12th are bad angles. Owner planets of these houses are considered bad. This becomes evident when stones are endorsed, and they too are placed in weak house i.e. in house below 80°. Instantly the symptoms aggravate but gradually becomes soothing within 3-4 days. This happens during stone therapy of bad angled weak planets. It is simply because the stone, as soon as endorsed, traps rays, by that time rays remain bad. Gradually it is altered by the RAD, i.e., stone and effects become soothing. Stones of good angle do not show such bad...to...good effects. They gradually and directly show good effects. Even stones of bad angled planets do not show bad to good effects if they are placed in powerful houses, some confluences like Jupiter–Moon, Sun–Mercury, Jupiter–Rahu, Rahu–Moon, Venus–Mars, Saturn in Libra, Jupiter in number of combinations, is remarkable.

However, 2 planets, 3 planets, 4 planets, 5 planets confluences effects are specific.

As already discussed, effects of Rays take place in three layers, covering very specific part in mechanism.

1. Saririk,.i.e, first layer (body physiology)
2. Atmik, i.e. second layer (body physiology and internal organs)
3. Anatmik, i.e. third layer (covering effects up to gene level)

First layer is covered by semiprecious singly or in combinations, sometimes work up to second upper (all body ailments). 2nd deep layer and third layers are covered by only precious stones. This is the site of fate i.e., genetic material (genes). So precious has alone power to effect deep layers, and tackle high frequency rays. Semi-precious cover broad spectrum and mixed rays but in combinations. Their combinations are very effective in required dozes and durations.

Trouble crops up in the individual due to the effect of bad Rays (Saturn, Rahu, Ketu's aspects planets periods) or bad angles of rays, i.e., period of owner planets of 3, 6, 7, 8, 12 houses. Periods of good planet, associated with bad planets, are regulated by endorsing dozes of combinations of semiprecious (if there is health problem). Precious may also be suggested in particular doze for recovery from recurrence and genetic nature of the disease. So, in the case of physical ailments70 to 100% recovery takes place. 50 - 60% recovery takes place in later phases of terminal disease and in case of genetic ailment too recovery is 50-70%, depending upon the 100% following up of directions. Of course, very few directions are given. Details of the treatments will be discussed on therapy site. Period (time) as told earlier is the main thing of one's life. This (time) may be called indirectly fate of the individual chance or sufferings of Individual.

Any individual who first steps in the cosmos switches on. Watch of his life since then he remains in constant control of the planets. This is to say that particular Individual's head comes out of mother's womb at particular time. Time refers to particular constellation, rashi and *rashi* of 1st house (Lagna) computed (process given). From the exact time period, constellation passed in the mother's womb and period remaining is computed. A constellation refers to particular planet So, the remaining period of that birth planet is the starting period of ones life. Then subsequent periods are calculated. This remains fixed for every individual. So, it is remarkable here that from the time and place of birth itself, individual gains or loses many things of life. Such as a person takes birth in period *Rahu*, his subsequent period will be Jupiter for 16 yrs, at the time of his career building. If Jupiter is of good angle and power person will go to any height in his career and Vise-versa.

Cycles of planets are fixed for all. Only there is lottery of nature regarding birth time, which not only gives good or bad periods rather good and bad confluences of planets in birth chart. Navmans and Gochar (present position of planets) are also computed to know one's present, future. So, time of birth and place of birth have great significance in one's life. Miracles of time has been elaborated. The place of birth signifies situation of planets and to some extent period also altitude, latitude. Minor change in place may not always matter.

Constellations	Planets and Periods
Pukhya, Anuradha, Uttarbhardara	Saturn 19 yrs
Ashlesha, Jyestha Rebati	Mercury 17 yrs
Varni, Mool Ashwani	Dragon's Tail (DT) 7 yrs
Poorva phalgun Poorva- kharha, Varni	Venus 20 yrs

The normal periods of planets are Jupiter...16 yrs, Mercury...17 yrs

Ketu...07 yrs, Venus...20 yrs, Sun...06 yrs, Moon...10 yrs, Mars...07 yrs

Rahu...18 yrs. A/C to bingsottari system prevailing in North India

Under one macro cycle of planet, periods of all planets rotate in cycle, e.g., in the period of Jupiter, micro cycle Jupiter falls, then micro cycle of Saturn, then according to above pattern microcycles of all planets pass in last *Rahu*. Thereafter, main period of Saturn, In one micro cycle all planets rotate in Sub micro cycle. So, there are 3 main cycles, Macro cycle, Micro cycle and sub micro cycle (*maha dasha, antardasha,* and *Prantyantar* dasha). It is common and given in books. Sub microcycle is not given in books. So, it can be easily calculated, e.g., In microcycles period of Jupiter all sub micro periods may be calculated on the formula. The periods of all three cycles are multiplied, and the total is divided by 40, gives the days. So, month and days will be calculated, that will be sub micro cycle (*pratyantar dasha*). Eg... we have to calculate JUP-JUP- Saturn, So, 16 x 16 x 19 = 4864/40 = 121.6 = days is Sub micro cycle of sat in JUP sub cycle.

Macro cycle (*mahadasha*) of the Individual is computed on the basis of Constellation. Every individual takes birth in a particular constellation; that too indicates a particular macro cycle. A part of that constellation is consumed in mother's womb and a part he/she enjoys after birth. So how much has expired before birth is calculated, and then balance period of constellation and hence macro cycle is calculated from balance calculation (*bhayat...bhavog, sadhana*). This *bhayat bhabhog sadhana* is available in all books. However, the main part of it which is not so commonly available is described as follows:

Constellations	**Macro Periods**
Kritika, Uttar phalgun, Uttarakharh	Sun 6 yrs
Rohini, Hast, sharvana	Moon 10 yrs
Mrigsira, Chitra, ghanistha	Mars 7 yrs
Ardara, Swati, Satvikha	Dragones Head (DH) 18 yrs
Punervasu, Visakha, Purva bhadra	Jupiter 16 yrs

(Incomplete chart, because it is common in Astrology books so, a pattern described). Scientific and statistical parts of astrology have only been followed.

Individual taking birth in particular Constellation undergoes macro period of upper mentioned planet. The periods of planet have already been mentioned. However, macro cycle at the time of birth reduces, since it is consumed in womb. Only balance is added in the beginning for further calculation. As told this balance of a macro planet is derived through *bhayat bhavog* calculation.

Balance period of macro cycle after birth is calculated in the following manners... . In cosmograph bhayat... bhavog are given such as... bhayat 16/39...bhavog 58/44

bhayat	bhabhog
16/39	58/44
x 60	x 60
960+39 = 999	3480+44 = 3524

If an individual takes birth in constellation *Kritika*, its owner planet is Sun whose macro period is 6 years. So,

999 x 6 = 5994 ÷ 3524= 1 year residual 2470 is multiplied by 12

2470 x 12 = 29640, the resultant again divided by bhabhog to give month

29640 ÷ 3524 = 8 months

Left 1448 x 30 = 43440 ÷ 3524 = 12 days. Approximately

1 yr. 8 months. 12 days (residual is multiplied by 30 and divided by *bhavog* to given days) This period is subtracted from 6 yrs....0...0...- Macro cycle of sun

= 1 8 12 days

4 yrs - 3m - 18 days approx.

For exact... 17 days...left x 60 ÷ 3524 = ghti

again left x 60 ÷ 3524 = pal. Hence, 4 yrs. 3 months 17 days 19 ghati 17 pal exact is balance of Sun macro cycle then 10 yrs moon...

So on so forth, as described previously...

In the midst of discussions one wonderful and interesting fact is disclosed. We all see that immense wealth is underneath in the different depths of the earth. It may be diamond, gold silver, platinum, iron, petroleum, innumerous kinds of ores and precious stones. Have u ever thought who has developed this enormous wealth? To our utter surprise, that these materials are extra territorial in origin. As already discussed, prior to organic evolution there occurred chemical evolution. This process has not stopped rather always going on. The Basis of this evolution is same, i.e., extra territorial cosmic ray and earlier mineral/ chemical of the earth. This transformation takes place at particular temperature and pressure with constant interaction of particular cosmic ray for thousands of years.

Cosmic rays in all such reactions are prominent, may be proved from the fact that U.S.A., thought to manufacture Diamond. For the same they developed all necessary conditions but failed to produce natural diamond, might be American Diamond. So, every factors they maneuvered, but failed to support energy of cosmic rays, constantly for thousands of years. Hence, they may be called extraterritorial in origin.

Relations of Cosmic Rays with Environment

Long ago, prior to the origin of life on earth i.e., during the days of chemical evolution, cosmic rays reacted with nonliving chemical/inorganic surfaces to produce several gases. Inadvertently among them oxygen and hydrogen were also produced which at certain temperature and pressure produced water on earth. This formed the basis of organic evolution. Then living forms started to develop. Again cosmic rays continued reacting with living, nonliving systems which developed a cyclic order of the environment.

For example, if photosynthesis alone is considered, which first of all results into formation of oxygen i.e. gaseous basis of life. The process of photosynthesis, when considered, it is the reaction between living chloroplast and non-living water, carbon dioxide in presence of solar energy. But the reaction finishes in dark phase when the rest part of cosmic rays (minus Sun), coming from stars, reacts and finishes the process. It is recalled in this connection that cosmic rays come from sun, stars, and galactic nebulae. Thus, in the formation of this basis in Environment (oxygen) Cosmic rays play vital role. The Sun is never alone is a source of cosmic rays.

Now it is evident that cosmic ray is the main factor in the development and evolution of plants on the Earth. These plants by evolving oxygen develop other life forms (animals). They in turn produce carbon dioxide, nitrogen and others for the formation of mixed wind.

Another main factor of the environment is water. This has also been considered developed with the vital mediation of cosmic rays. It is assumed that at the time of geological development of earth, there was Chemical/inorganic surface and cosmic rays alone and there developed water. This is just possible that at the time of separation of planets, earth shared such types of chemicals which so ever under reaction with cosmic rays produced gases like the main oxygen and hydrogen which through bond formation (utilizing same cosmic energy) produced first water molecules that in turn in a cyclic order became the basis of life and the environment. Probably such phenomenon did not take place on any planet other than earth (May be due to its chemical/mineral composition) Consequently, it is not at all exaggeration to proclaim that Cosmic rays also played vital role right from origin to development of different constituents of environment on the planet Earth.

Cosmic Rays and Precious, Semi Precious Stones

Slight discussion has already been done in this regard, as it is the main part of Health n wellness, so detailed. Cosmic rays constantly falling on the earth piece, react with chemical/ mineral part of the earth at any depth for thousands of years. That particular part of the earth undergoes chemical evolution becomes rays absorbing or altering. So, they may be called RADS (Rays absorbing/altering Device). But when rays reacting with particular part up to shorter period say for hundreds of years, they become semi-precious. For example, if rays of Jupiter characteristics falling on the part of Earth for thousands of years that part (mines) develop as Sapphire. Due to difference in periods they become white another Yellow. In this way, when a part of chemical/mineral becomes exposed for hundreds of years that part is converted into Semi precious (Topaz). Likewise all precious and semi-precious under the control of each types of pure rays develop as diamond, emerald, blue sapphire, ruby, Yellow sapphire, cats eye, precious hessonite and corals and pearls (in sea). There are numerous types of semi-precious stones that are developed after respectively short exposition. For therapy, following semi-precious were selected depending upon the apparent efficacy corresponding to dozes. Bead, preparation with whole facilities for every semi-precious coral, garnet, topaz, lepis, hessonite, tiger eye, cats eye, crystals, American diamond and sea pearl, emerald beads, phiroja, etc.

Rays falling on earth are not angle specific. They are always of good angle. Every individual is taking birth at a particular time and place forming angles with every type of rays. Among the angles few of them are good and few are bad. While some of them are very lethal for example, 2, 3, 6, 7, 8 and 12 th angles are bad, among them 6, 8 and 12th are lethal. The stones of these rays react or aggrevate the symptoms for few days then become converted to good effects (i.e., bad to good). It is simply because, as soon as the stone converts the rays, it is taken in the body in bad angle, aggravating the symptoms. So, stones take few hours to change angles. This bad to good effects are only achieved during regulations of bad angled rays. Sometimes lethal rays become hard to be regulated. Then recourse with 90d planet is conducted or the concentration of rays of good planet is increased. Aggravation of symptoms is the evidence of bad to good conversion of angles.

(Parts of sky) Approximate 30° each (360/12) (Diagrammatic)

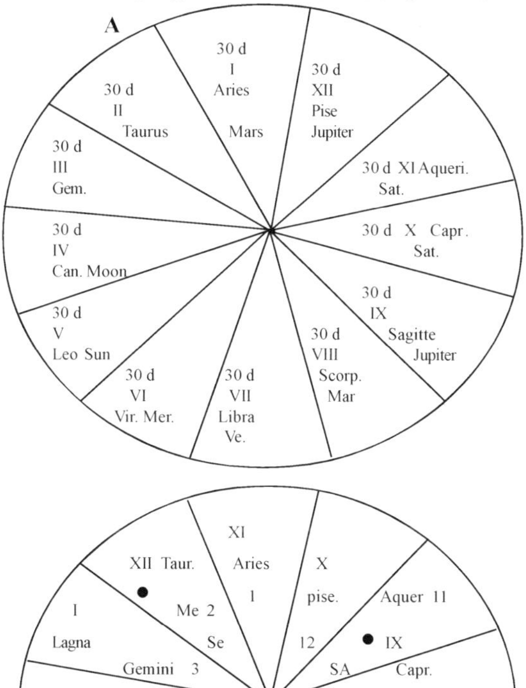

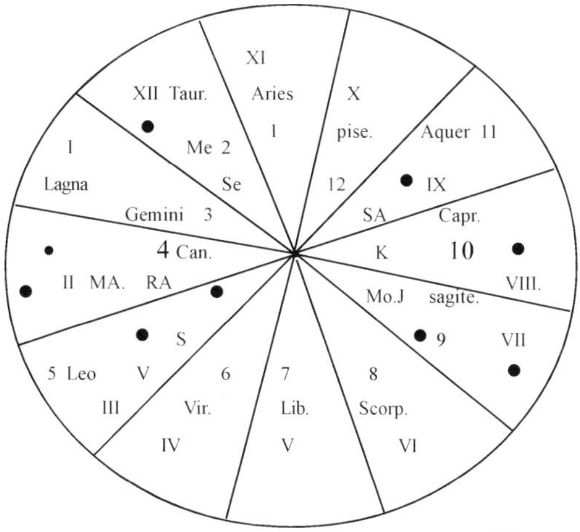

Horoscope/cosmograph (B) **Diagrammatic**

Fig A Reveals different parts of the sky (also discussed earlier). There are 12 parts each of 30 d... called *Rashies*, consisting of stars of more less same characteristics. This is bigger group of stars, but first unit/tool for characterization of cosmic Rays (which are main actors behind the full show). For minute details of stars the whole sky was divided into 27 parts. Each one was named contellation. Here we are concerned with 12 parts as *rashies*.

1st...30d...is...Aries...owner...Mars...Colour...orange

2nd...30d...is...Taurus...owner...Venus...Colour...white

3rd...30d...is...Gemini...owner...Murcury...Colour...green

4th...30d...is...Cancer...owner...Moon...Colour...mercury

5th...30d...is...Leo...owner...Sun...Colour...red

6th...30d...is...Virgo...owner...Mercury...Colour...green

7th...30d...is...Libra...owner...Venus...Colour...white

8th...30d...is...Scorpio...owner...Mars...Colour...orange

9th...30d...is...Sagittarious...owner...Jupiter...Colour...yellow

10th...30d...is...Capricorn...owner...Saturn...Colour...blue

11th...30d...is...Aquarius...owner...Saturn...Colour...sky

12th...30d...is...Pisces...owner...Jupiter...Colour...yellow

On the platform of Astrogenetics it was observed that both parts of the sky (houses) have impacts, i.e., general house and individuals house. If both are handled and regulated then full effect of that house is achieved. For example, if an individual requires regulation of 4th house then his own 4th (4th from Lagna) and moon are regulated to achieve full result of 4th house of the person. So, this fig. A is very important and universally required. The power is only acquired through personal cosmograph i.e., power of the planets are calculated from one's own 1st (Lagna) house 1, 4, 7, 10, 5, 9 are powerful houses, and 3, 6, 8, 11, 12 are weak houses from ones Lagna (1st house). Dominance and recessive factors will also be determined from individual's 1st. house not general ones. So, factors (planets) present in 1, 4, 7, 10, 9, 5 houses are Dominant while those present in 3, 6, 8, 12 are recessive. Those present in 2nd are dominant Lethal. As told 6, 8, 12 are recessive lethal;

except Venus in second i.e. for beauty, and Saturn in 8th... for longevity, in addition Saturn in 6th... i.e., antidote of enemy. Moreover, factorial display will start from fig. B, of course mock.

Different parts of Sky, their owners, colours; each of 30d are arranged serially above.

Characteristics of houses have been discussed in previous pages. These are general arrangements, for all persons. But by virtue of Ones place of birth and time any number within 12 may be calculated as Lagna or 1st house of the individual. So, a second cycle, first is general and second of a particular individual rarely one when Aries is 1st and Lagna both of an individual. It becomes more evident from the following example:-

This Fig. B also has 12 parts but as an example... 3rd is 1st here. Meaning there by according to the time of birth the first house was calculated (Lagna), as Gemini. So, 3rd becomes the 1st of the individual. This is individual difference. Hence, everything is changed, e.g., Personality, brain, etc. they are being regulated by mercury, the owner of the house, wealth from Mo. and so on fig. B...some planets have also been placed, showing the mode of interactions. However, under total interactions:

27 x 27 x 12 x 12 x 9 x 9 = 8,503,056 factors/genes

i.e., Constellations x constellations x rashi x rashis x planet x planet (x in multiplication).

But broadly *rashies*, planets interactions are apparent. In Fig. 3rd (personal) house is showing interaction of Sun and Venus i.e. the owner of 3rd is Sun, which interacts with the planet Venus. That may be called as sun part of rays after taking characters of Venus (reflecting) regulates 3rd part of fate and organs. The owner of 4th (P) part is mercury. So, characteristics of 4th house (part of sky) is regulated by Mercury. Similarly 6th house is regulated by orange coloured Mars. 5th house (part of sky) is regulated directly with Venus. 7th part is regulated by Jupiter through Jupiter and Moon. 8th part is regulated by Saturn through Ketu. 9th part is regulated by Saturn. 10th by Jupiter through Mercury. 11th by Mars. But again 12th is regulated by Venus through Mercury. These pure or interacted rays are energy support and regulate development, during differentiation, and functions of organs, physiology, and genes.

That's why genetic functions are also reflected which is a matter of great importance for Geneticists.

So, in fig. B Jupiter & Moon are dominant, as 7th is powerful house and also Jupiter is in double doze. Thus, the characters linked with Jupiter are dominant.

Characters linked with 3rd (v) are recessive but (s) are dominant. Characters linked with 7th and regulated through Jupiter, Moon interaction are partly dominant. Again characters of 9, 10 are dominant regulated by Sat. and Jup. characters linked with 8th regulated by Saturn are recessive. The owner of 8th i.e., Sat, placed in IXth is dominant itself becoming positive in some aspects but putting bad effects to 12th. Here it is to be mentioned that bad forces when weak considered positive.

Special situation is described by saying, sat. is good because when owner of bad becomes situated in bad house gives some good effects. Bad and good planets are Sat, Rah, Ketu, Mars, Sun are considered bad planets but under some circumstances becomes also good. Mercury, Venus, moon, Jupiter are considered good but owners of 3, 6, 8, 12, and 2 are bad. Moreover we have least concern with good or bad as we have to explain genetic behaviours, and inheritances. For readers interest something regarding good or bad have been elaborated and range of mechanism illustrated. In this way dominant and recessive characters have been illustrated in brief.

Relation of Colour and Cosmic Rays Spectrum

Scientific view of Spectrum reveals that it consists of 7 colours but according to astrogenetic observations 9 colours have been reported. Besides seven, white and mercury are also included, white is the base of all colours and mercury is of moon (reflected). These Nine, rays appear more exact on rotation of the spectrum disk. The effect of colours is also visible in the therapy. The stones under regulation are of different colours. The difference in colours deepness affect differently. In the spectrum, only red part is of Sun, others coming from stars of different part of sky, as already reported. As such,

- Colour of 1 and 8 parts of sky gives orange.
- Colour 2 and 7 parts. white, 3 and 6 part green

9 and 12 parts yellow, 4th part – Mercury, 5th part – Red, 10 & 11 th parts blue and sky colours respectively. Sky parts may be calculated from Equator. 15 d right and 15 d Left of the hypothetical line of the Equator is 1st part of sky of 30 d; then each 30 d is 2nd, 3rd, 4th and so on so forth. The attributes of the colours are, important hence colour therapy is also effective a bit in the range of 5 _ 10%. So, semiprecious stones are required in deep colours while colours in precious make them different in Nature. They become specific for certain patients as white saphire cannot be suitable for yellow requiring patient. So, colour requirement varies from patient to patient. Here difference in deepness of colours becomes directly proportional to effects.

Living and non-living activity is also related in Nature through cosmic rays. (discussed in the aura mechanism).

This is Specific of astrogenetics may be confirmed through kirlion photography.

Ionospheres

It is in the way of cosmic rays coming from Stars to the earth surface and extended in the atmosphere above 50 to 400 miles. It reflects electromagnetic radiations. The rays ionize the materials present in the atmosphere and they remain in ionic state in atmosphere. The atmosphere in ionic condition is called ionosphere. This ionosphere becomes the good conductor of electricity because free electrons and positive ions remain present there. It also acts as reflecting surface of radio waves. Upper surface of atmosphere develops increase in temperature due to ultraviolet radiations of Sun. It is also ionized to greater extent.

Constellations are divided into 4 stages, and each signifies a letter 9 stages = 1 Rashi

Constellation	**Stages. Letters.**	**Rashis**
Aswani	chu, che, cho, LA	Aries
Bharni	Lee, Lu, Le, Lo	do
Kritika	Aa	do
Kritika	ee, u, ye	Taurus
Rohani	o, ba, bi, bu	do
Mrigsira	be, bo	do
Mrigsira	ka, ki	Gemini
Ardra	ka, gha, unga, chh	do
Punervasu	ke, ko, ha	Gemini
Punervasu	hee	Cancer
Pukhya	hop, he, ho, da	do
Aaslekha	dee, doo, de, do	do
Magha	bha, bhi, mu, ma	Leo
Puurvaphalguni	bho, ta, tee, too	do
Uttarphalguni	Te	do
Uttarphalguni	to, pa, pi	Virgo
Hast	pu, pa, ara, dha	do
Chitra	pe, po	do
Chitra	ra, ree	Libra
Swati	ru, re, ro, ta	do
Vishakha	tee, tu, te	do
Vishakha	to	Scorpio
Anuradha	na, no, nu, be	do
Jyestha	no, ya, ye, yu	do
Mool	ye, yo, bha, bhi	Sagittarius
Purvakharh	bhu, dha, ph, da	do
Uttarakharh	bhe	Sagittarius
Uttarakharh	bho, ja, ji	Capriccorn
Shravan	khi, khu, khe, kho	do

Ghanistha	ga, gi	do
Ghanistha	gu, ge	Aquarius
Satvikha	go, sa, si, soo	do
Purvabhardra	se, so, da	do
Purvabhardra	dee	Pisces
Uttarbhardra	doo, tha, bha, yan	do
Rebti	de, do, cha, chhi	do

Persons take birth at a particular time and place. Their constellations are worked out. The stage of the constellations are computed (through *bhayat bhavog* calculation). From the above chart stage wise letter is derived. This letter should be the beginning letter of one's cosmographic (rashi Name). Where the moon is situated in the cosmograph. That rashi is called individual rashi. The rashi becomes the first house of the present planetary status and effects i.e. (Gochar effects). So,

1. Rashi
2. Constellation
3. Moon cosmograph

These are helpful to know the present status of the individual Sometimes, Navmangs Cosmograph is also required for the purpose. However, they are required for predictions.

Fig. A ... Presents the views of all 12 parts of sky/houses, their Colours, and attributes. The dominant and recessives are also decided according to the situation of regulating planet. If it is situated in powerful house e.g., 1, 4, 7, 10, 5, 9. .... it is dominant otherwise. Recessive, some are recessive lethal, i.e., owners of 6, 8, and 12th. placed in weak house. Owner of 2nd when placed in 6, 8, 12, is Recessive lethal given in Fig. B.

Rays coming from stars act on genes, reactive surface of cosmograph found acting as genome, Since they take part in genetic processes i.e. more emphatically the plain surfaces of Cosmograph with interactions found Genome activating because they exhibit genetic phenomenon. So, 30° each part of it may be gene/genes of specified characters and under control of stars of rashis and constellations. So, they are inheritable and rays coming from stars after interaction with planets interact with genes. They are also inheritable, may be or may not be in every offspring in every houses or part of genome.

Reflections of Planets

Although this has already been illustrated, but for exploring and deciding interactions, it is essential to discuss. Every planet reflects towards 3, 5, 7 houses, Mars 4th, 8th also Jupiter 3, 5, 7, 9 and Saturn 3, 5, 7, 10 (verified by A.G. during therapy).

AS in fig. B:

1. Factor Me interacts with MA, RA of Can.
2. MA, RA interact with K of Capr.
3. S and V of Leo interact with Mo, J factor of Sagit. and SA of Aquer.
4. Mo, J of Sagit. Interact with SA of Aquer.

Besides J of VIIth house (Sagit.) interacts with V and S of III (Leo) and SA of (IXth) Aqu. Now, each interaction illustrated may be well interpreted i e. consisting of many interactions of geometric parts for enclosed traits or genes.

Aa in 1st (a) Me interacts with XII house (b) Me interacts with organs and genes of taur (c) Me - MA interaction (d) Me - RA interaction (e) MA x RA x Me interactions. Those interactions are effective on human organs.

FATE - inheritance will be worked out in cosmographs arranged for two, three generations. If these rays are TRUE vehicles for the transmission of characters from one generation to the other, the inheritance of characters will appear from both and Genetic plays will be observed in real chart of cosmographs. (cf. figs. A2 B2 C2).

- Inheritance of factor (Rasies, Planets interactions)
- Justifying, astrogenetics Nomenclature

EXAMPLE 1... Figs A2, B2, C2

From aforesaid it is apparent that Constellations, Rashies and Planets, through Cosmic Rays activate human genome (characters/genes). Such belief (hypothesis) is being substantiated scientifically in real Cosmographs of two generations constructed at different (long spans of times). The process of formation, Computation of cosmograph is same as that of the Horoscope in old science. Few changes have been done on the basis of observations made during 25 - 30 yrs of active experiments.

The process of computation will be discussed later on. In figs. A2 is mother, B2 is father and C2is only daughter.

A2... M...	F... B2....	D.... C2...
L... 1st house	L... 1st house	L... 1st house
8. no. orange	9. no. with full	3. no. in which
colour, owner	impact of MA	MA is
Mars. MA. (f)	due to presence	present 3 no.
90 power	in 7th, power 90.	power 90 D.

In both the parents, the MA factor is very prominent and dominant in L (first house) it is also present in L of daughter. Is it by chance? A2 cosmograph was prepared on 28. 4. 81, B2, prepared on 11. 9. 77 and C2 prepared on 9. 11. 2007. More over this family was randomly selected. So, orange coloured Cosmic rays MA activating head, brain genes being dominant in both parents, Inherited in daughter; controlling same factors. So, it is seriously suspected that cosmic rays (MA) acted on genetic transmission from parent to offspring in dominant conditions.

2nd Factor :-

R...	R...	R...
It is present in	It is present	It is present
9th part in M.	in 9th part. F.	in 9th part of D
Power being 100	Power 100 D.	Power 100 D.

Clear example of inheritance of dominant character.

Dominant in both parents, hence transmitted, inheritance of SU - ME.

A2	B2	C2
This is present but in 6th (weak) power 60.	It is present in 9th. Where SU in Power 100.	Present in 5th, Power 100.

SU ME is an intellectual character, which is present in both parents, weak in mother, powerful in father hence inherited from father, mother simply helped Father. The father is learned Engineer but mother is house wife. Daughter has brilliant career.

Photos exhibiting inheritance of factors/genes in real cosmographs prepared at different times noted below. They are of mother, father and daughter. Dominant, Recessive characters are very well conceivable and perceptible, those up till now were not so. If same factors are transferred to offspring in more or less same pattern tells the story of transmission of characters (inheritance). So, it is justifying the nomenclature.

ASTROGENETICS EXAMPLE 1

Cosmographs Randomly selected Pedigree 1

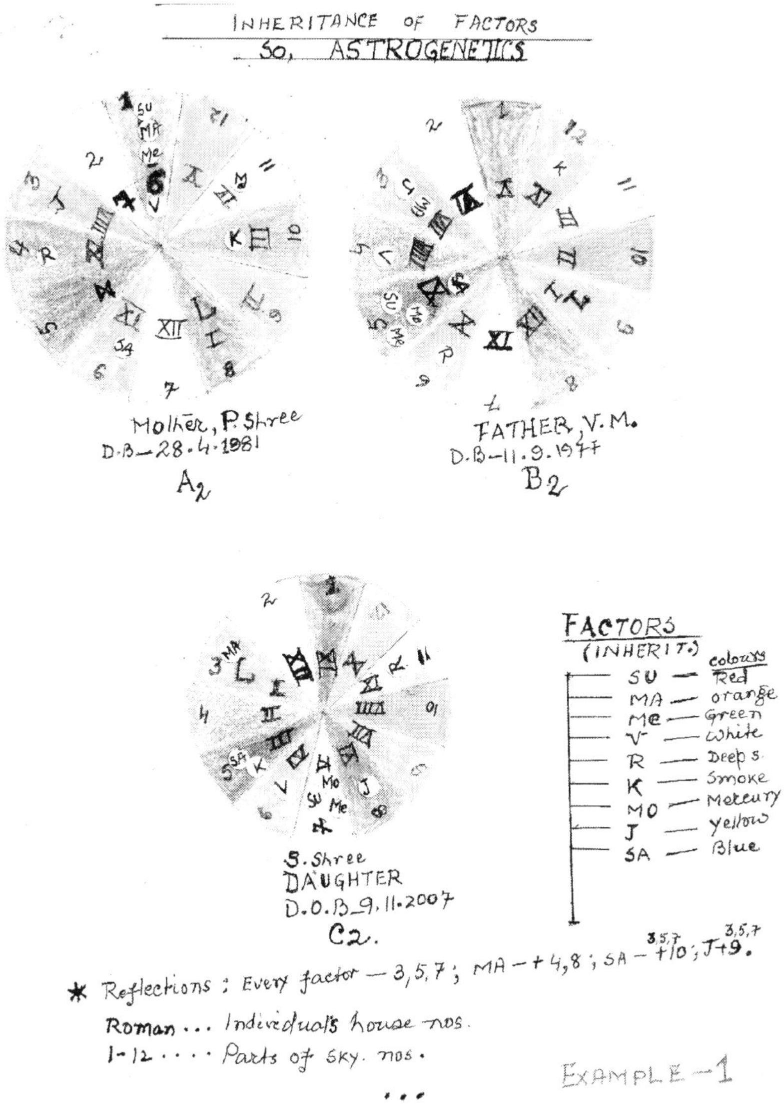

Mother, P, shree
D.B. - 28.4.1981
A2.

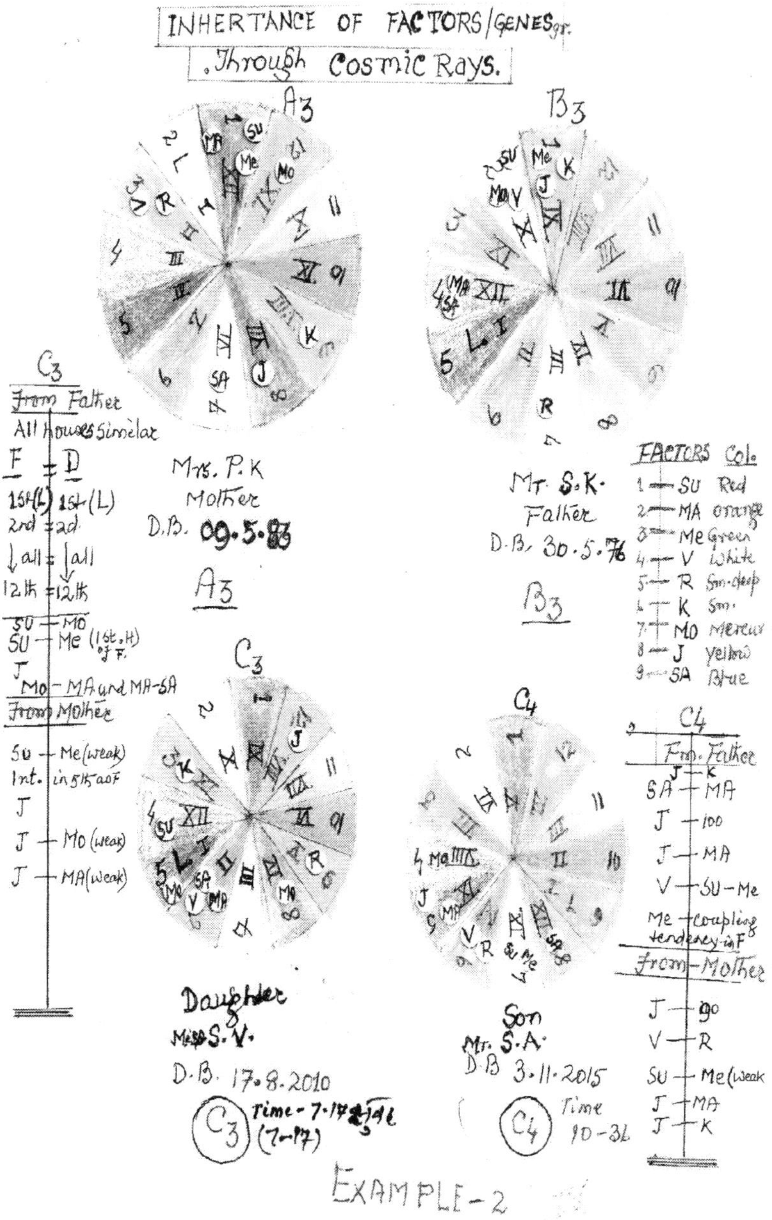

SU...Me. From father, supported by mother, dominant in father, recessive in mother.

R...from mother supported by father, since 9th in mother and daughter, 10th in father, 100; 100 power in mother daughter but 90 in father, all dominant.

INHERITANCE OF FACTORS/GENES
In the Real Family Cosmographs.

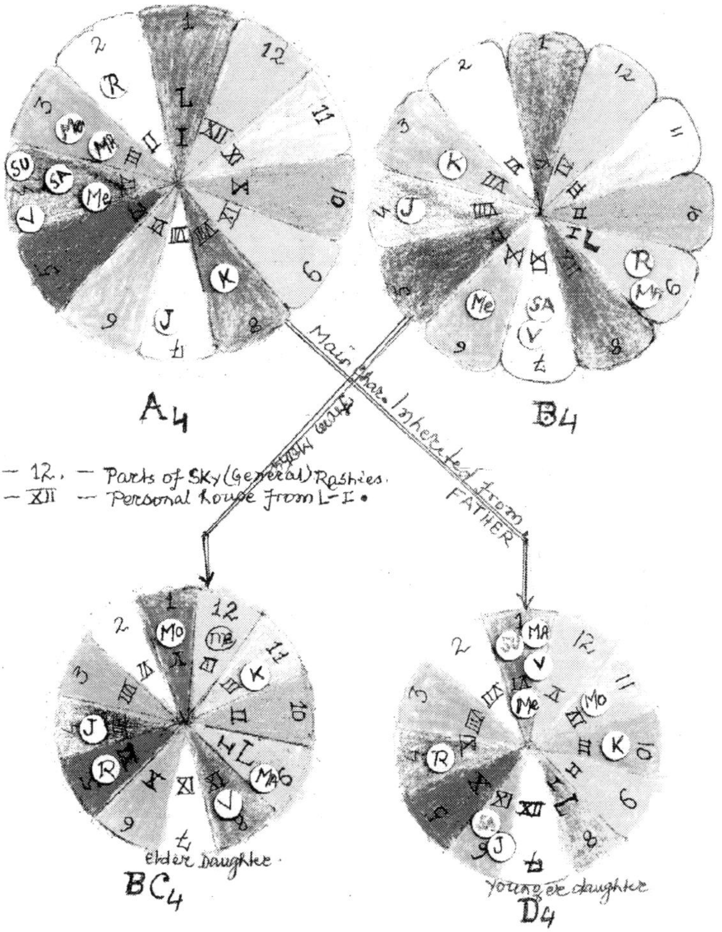

EXAMPLE —3.

* A4 — Father ; B4 — Mother, BC4 — Daugher ; D4 - Daughter
D.B — 13.8.47 D.B. 21.10.1954 D.B. 8.1.79 I D.B - 29.4.81 II

EXAMPLE - IV

80

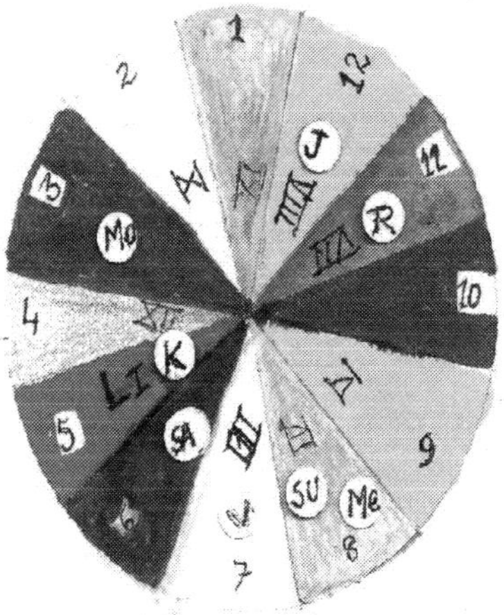

A Father -(D.B. 13.12.51)

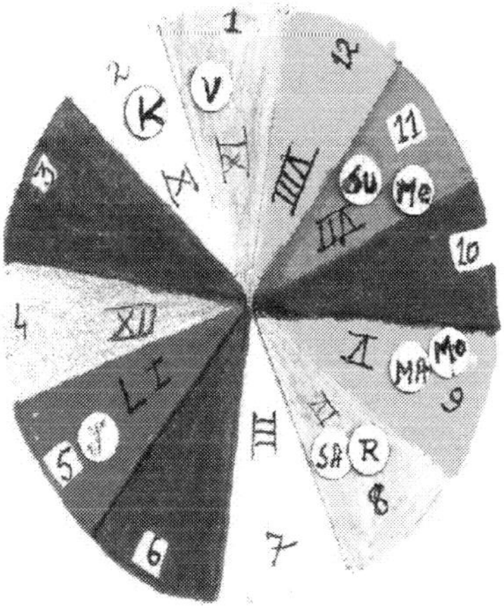

B Mother
(22.9.55)

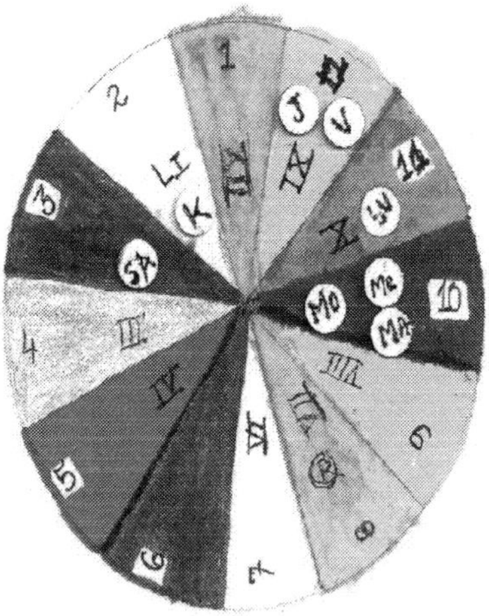

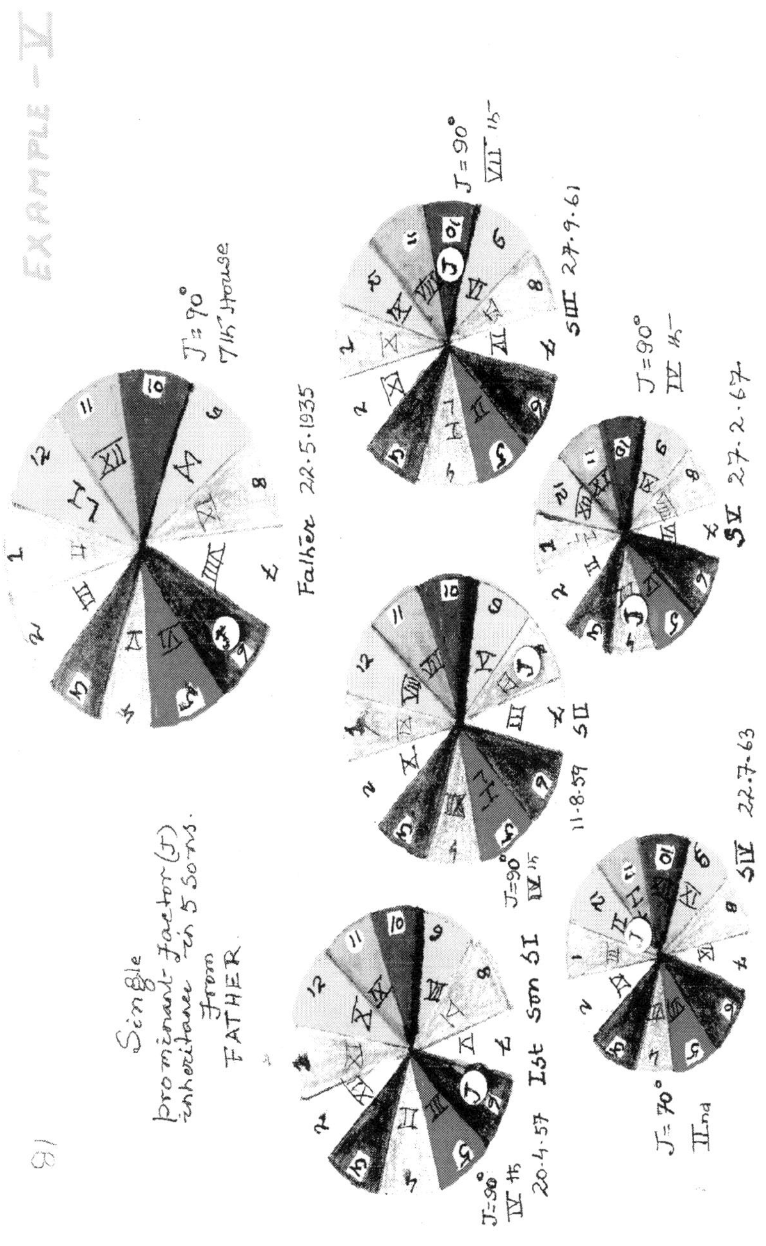
EXAMPLE - V
Single
prominant factor (J)
inheritance in 5 sons.
from
FATHER
J=90°
7th House
Father 22.5.1935
J=90°
VII th
S III 27.9.61
J=90°
IV th
S V 27.2.67
11.8.59
S II
J=90°
IV th
S IV 22.7.63
J=70°
IInd
J=90°
IV th
20.4.57
Ist Son SI

Inheritance of. Su - Me - Mo

These set of genes are inherited in daughter from father. In father they are dominant, present in 9th house (power 100), in mother a. separate interactions (Su, Me, MA, V) is present but recessive. So, dominant interactions from 9th inherited in powerful house 7th of daughter, where also they are dominant. Mother too supported, even if it is recessive but interaction is present.

MA - J... inheritance

This interaction is present in father but not in mother. Not only that, in mother MA, and J both are recessive placed in 6th and 11th Houses (power 60). In father they are in 9th (power 100).... so from father these dominant features inherited in daughters 5th (power 100).

Interaction V - Mo inheritance (MO is the house)

Above interaction developed in Father, but it is not present in mother, even in father this is weak ie in 8th (60). But this interaction inherited in daughter.

MA - Me... interaction inheritance

This interaction is present in both the parents but weak i.e., in 6th and 8th respectively. However it appeared in daughter as dominant feature i.e., in L (first). In nutshell, Factor MA, contributed by mother (as owner of the 1st house and 8th impact on L). Father supports i.e., through direct impact on L(1st).

SUMe from father, supported by mother; dominant in father, recessive in mother.

R... from mother supported by father, since 9th in mother and daughter, 10th in father 90;. 100, power in mother daughter but 90 in father, all dominant.

EXAMPLE 2nd of real cosmograph

Inheritance of factors/genes

....

A3... is Mother... D.B. .... 9.5.1983...

B3... is Father... D.B. .... 30.5.1976...

C3...is Daughter...D.B. .... 17.8.2010...

C4...is Son ... D.B. .... 3.11.2015...

....

Inheritance of 5th part of the sky, SU from father A3 to daughter C3 from 1st House. to. 1st house (red colour is SU in both).

So, all houses of D...C3, inherited. Similarly, from F A3 i.e., personal no. of house rt from L I to XII. Apparently, colours of L I to L XII may be observed similar in colours ie. regulator rays (Rashis) of both equal. It means regulating rays of father and daughter are similar. A good example of inheritance of gross factors from father (B3) to daughter (C3) Displayed.

SU...MO

It is inherited to daughter from father 10th H. In daughter SU is present in 4 (part of sky, house) MO in 8th, forming interaction in 8th. Mercury Su of daughter placed in the house of MO.

SU Me

In father, above interaction develops in L, when Me present in IXth; but powerful... 100° p. In mother it is in XIIth house, weak 60. So, it is inherited in Lof D. (Red is SU and Me present) C3 from father with the help of mother.

FACTOR....J...

In both the parents this factor is powerful and Dominant. In F B3, it is present in personal 9th, in M A3, it is present in personal 7th Personal means no. of house from L, mentioned in Roman. So, it is inherited in daughter, of course in 8th, but in double power i.e., it is situated in same house J(in the house of J). In C4, Son L I and IV th houses are regulated by J.

MO MA

In father this interaction develops in 12th personal while in mother it is absent, so it is inherited in daughter from father alone. In daughter C3 it is in powerful house 4th personal. p90.

It is perhaps, in mother Mo...MA has coupling tendency in 11th and 12th, personal. So, they might have helped in becoming dominant in daughter C3.

MA SA

This interaction develops in MA3, in 6th personal. MA is in 12th, SA in 6th. So, develops in mother in 6th, personal. Hence, it is weak in mother. In father also, it is present in 12th personal, i.e., weak. Since, both the parents have contributed, so it became slight powerful by becoming situated in 2nd personal in C3. This is a bit bad house. While viewing interactions, colour of the house and planet situated both are taken into account. Colour of the house is main regulating force (Rashi) and planet is interacting, reflecting object.

J Mo

This interaction develops in 11th personal in M...A3, when this interaction is not present in F. B3 but has a coupling tendency in powerful houses 9, 10th personal. This interaction develops in 4th, personal house of daughter. Mo in 4th of D, C3 and J in 8th personal, so interaction with 9th aspect of Jupiter takes place in 4th personal of C3.

J.... MA

The above interaction is present in 7th personal of M, A3. This develops in F, B3 in 9th personal. This is inherited in 2nd Personal of daughter, slightly less in power but wrong. This develops In D, C3... due to presence of J in 8th pers. and MA in 2nd. In C4 Son this interaction is present in XIth but weak.

SA...MA

This interaction is present in F, B3 in 12th personal and inherited in son, C_4 in 12th personal. In mother also, this develops in 12th personal. This interaction is present in all three in 12th except D, C3.

J....K

In F, B3 it is present in 9th personal p100 and in M, A3 it is in 8th personal p60. It is inherited in 4th personal p90 of C Son.

Hence, this interaction is inherited from father supported by mother. Daughter failed to inherit, perhaps because it is weak in mother, might be sex linked inheritance.

Inheritance of J. in son it is in IXth and regulates L I & IV P. In F. B3, it is present in 9th personal. p100. In mother A3, it is present in 7th personal p90. So, this is present in both the parents as Dominant. Hence inherited in son in 9th p100. As discussed it is also inherited in D, C3, of course in weak house but in double doze i.e., J in the house of J. but in 8th.

V...SU...Me

This interaction develops in 6th personal of M, A3 by interaction. with 12th personal Power 60 only. Where as in F, B3 this is likely to develop in 10th personal, as Me; from 9th has a tendency to couple Pertaining to the fact that it is inherited in 11th personal of son, C4. P60, but Good.

SU...Me, as already discussed is also inherited in Son, C4 in 11th personal. It is present in mother in 12th personal and in father it is. Present in LI personal p90. So, it is inherited from father, mother supported in son.

V....R

This interaction is for allergic manifestation, Present in M, A3, in 2nd, F, B3 it has developed in 3rd, power 60. The power is slightly greater. In M, A3 i.e., p70 is bad. It is inherited in son in 10th personal, perhaps from mother. He developed allergic manifestations right from birth. So, he was kept in incubator with oxygen for 48 hrs.

Cosmographs showing inheritance of factors/genes in a Real family A3 - Mother, B3 - Father, C3 - Daughter, C4 - Son is a testimony for live study. The inheritance of factor by the mediation of cosmic rays further strengthens the view that cosmic rays are causes of Genetic phenomenon and whole interaction surface of Cosmograph is genome/genes regulating forces of the individual bearing that Cosmograph.

COSMOGRAPHS showing inheritance of factors/genes in a Real family. A3 _Mother, B3 _Father, C3 _Daughter, C4 _Son.
(Diagrammatic).

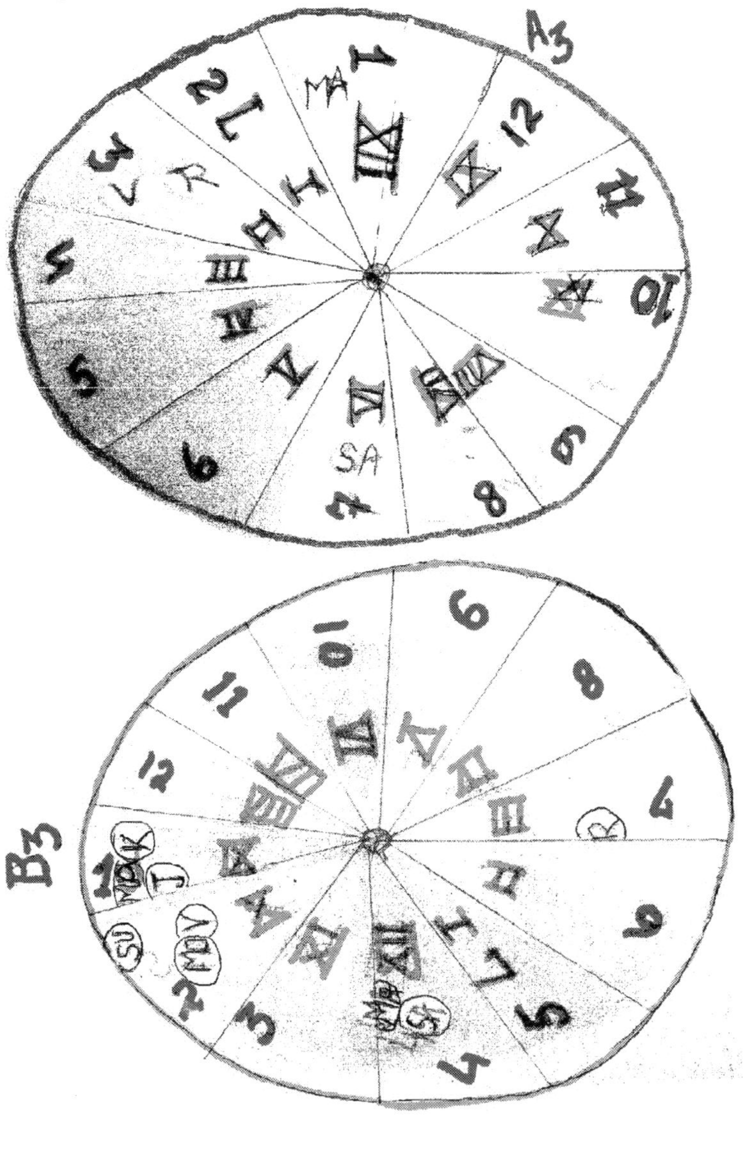

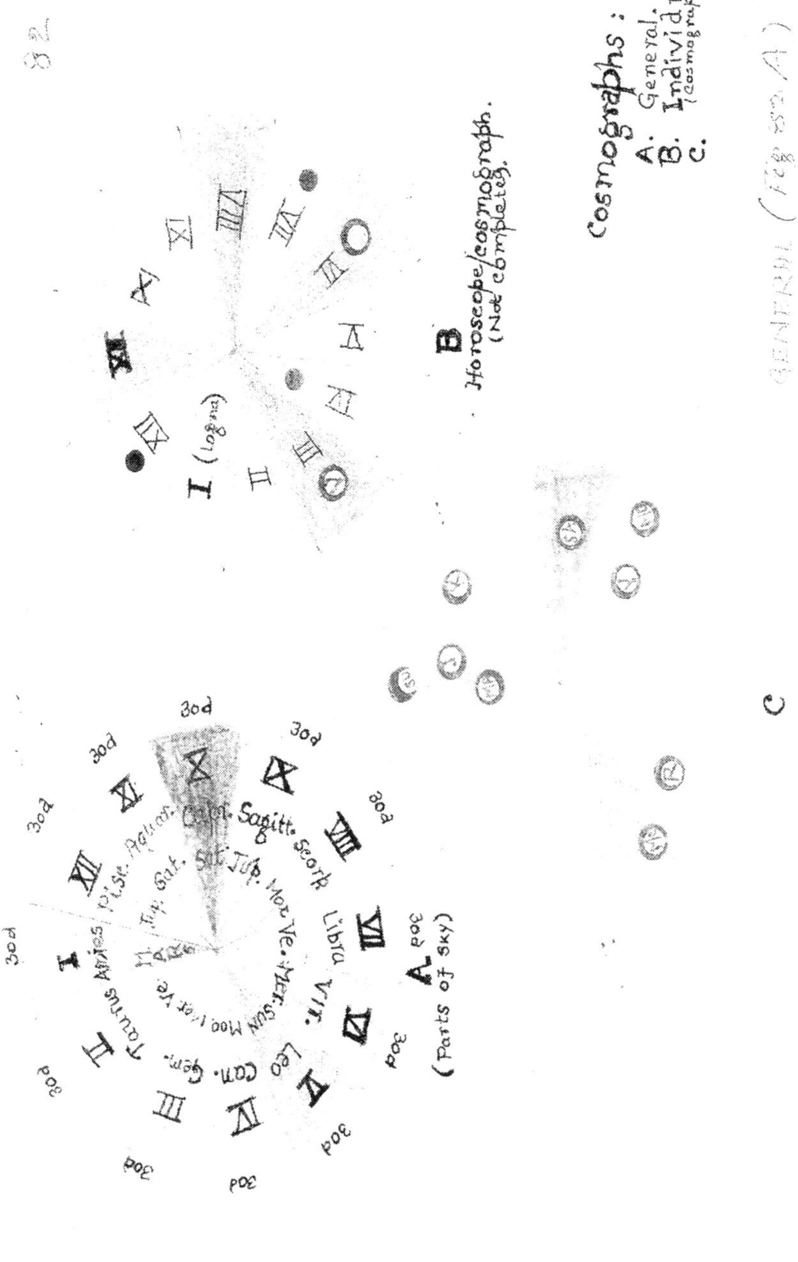
Cosmographs :
A. General.
B. Individual (cosmograph)
C.
B Horoscope/cosmograph. (Not completed)
A (Parts of sky)
C

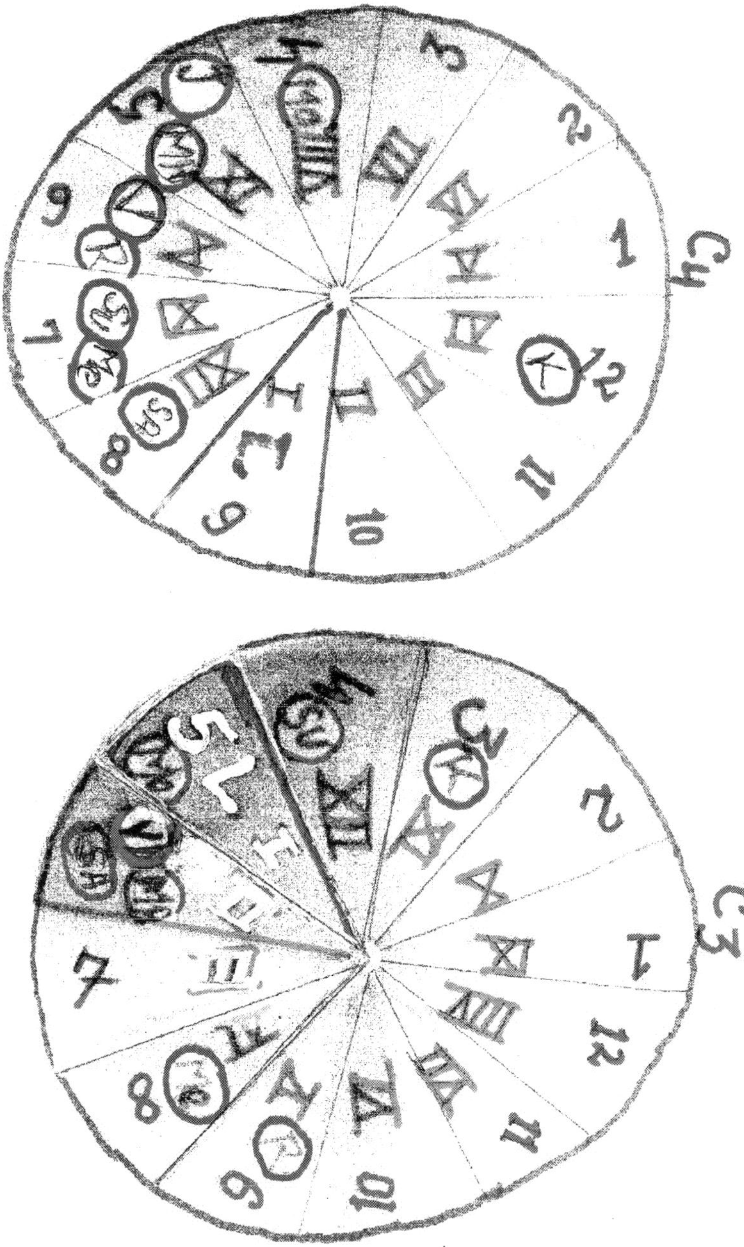

(Diagrammatic)

Cosmograph preparation is statistical computing but based on Indian Astrology under compromise in illustrating Astrological/Scientific scrutiny in Hindi.

There are five facts which define:-

Astrology in simple way...

1/Hora (Ahoratri),

2/statistics,

3/Sanghita,

4/Questions,

5/Nimitta (concerning).

1/Hora is abbreviation of Ahoratra, by deleting A, and. tra; means individual who takes birth. This hora explains aspects of planets in relation to time and place of birth of individual. Time signifies planetary situation of the moment and place denotes altitude & latitude. So, this chapter deals Hindi/Sanskrit terms. Some are translated but some of them in the same form i.e., in Hindi or Sanskrit but with explanations for comprehensive description.

Thus people have to follow, however very simple presentation and statistics used for all viewers. The pointwise brief presentation is as following... 1st kaal/ist dand...it is calculated from time of birth (first oozing out of head, or just bringing out in. Scissarian i.e., first appearance time in this cosmos).

The time of birth is converted into Astrological units time i.e., Dand and Pal. statistics-hour-minute of birth time. It is corrected with Desantar and Belantar (given in charts). So, corrected hour- minute, is deducted from time of sunrise hour-minute.

The resultant is multiplied by 2.5

(Birth time (± Desantar, Belantar) Sunrise)

x 2.5

Note if birth time falls between 12 noon to 12 night, 12 hours is added in birth time. If birth time falls between 12 night to next sunrise, 24 hours be added in birth time.

Above corrections in birth time is done simply because standard time of the whole country is same, but dhup gharee (solar watch) time is variable. Thus birth time is corrected for variable place of birth (longitude)/latitude...rekhans/akchhans and standard/local time difference is noted.

Example... of above statistics...1/

30/8/2016 ..... Birth time 9am

Birthplace..........darbhanga,

Sunrise............5/41

Desanter of darbhanga...........+13 minute 48 seconds.

Belanter of 30 August...........- (1minute)

So, following rules, apply:

Hour	:	Minute	:	Second
9	:	0	:	0
		13	:	48 — 1
9	:	12	:	48
-5	:	41	:	0
3	:	31	:	48 x 2.5
7	:	30		
1	:	17	:	30
+0	:	2	:	0
8	:	49	:	30

= 8 dand and 50 Pal.

Example...II

II. Twelve hours are added if birth time is 6 pm.

Hours Minutes Seconds

6 00.00

+12.	00.	00
+00.	13.	48
-00.	1.	00
18.	12.	48
-5.	41.	00

12	31.	48
x		2.5
31.	19.	30

= 31 dand. 20 pal.

Example III

If birth time is 2 am night, 24 hours are added.

Hrs.... Minutes.... Seconds

2. 00. 00

+24.	00.	00
+00.	13.	48
-00.	1.	00
26.	12.	48
-5.	41.	00
20.	31.	48
	x	2.5
51.	19.	30

= 51 dand 20 pal.

1. If birth takes place after sun set to 12 night the above resultant is added in din maan toget 1st kaal.
2. If birth takes place between 12 noon to Sun set the resultant is deducted from Din maan to set ist kaal.

Lagna in horoscope, L in cosmograoh or house of individuals horoscope is determined as following :-

According to example 1st the birth day is 30/8/2016. Sun rashi degree (ie aspastsurya). is taken from *Panchang* of 30/8/2016.

The value indicated is...4/13/6/11. This means Sun has entered Leo (Singh rashi) at 13 degree. Now from Lagna table the value of 13 degree Leo (13 degree per) (Singh lagna ka man) is taken. It is 25 dand 30 pal (Singh Lagna period) Now. 1st dand is added.

Dand	Pal
25	30
+ 8	50
34	20

This 34 dand 20 Pal is found in (Tula rashi) between 0 to 1 degree. So, Lagna or 1st house or L of cosmography is in (Tula), rashi (at 0 degree). It is presented as 7/0/0/0.

Thus Lagna/1st house/L. is 7/00/00/00/.

Planetary Position

After determining Lagna, birth chart is prepared where Lagna/L is indicated as 1st house personal/L on the top and subsequently in anticlockwise manner other rasies are indicated; Here Lagna is Libra, so it is indicated at the top (first house), and other rashies gradually on left side. Then according to panchang (planetery position book) of that day planetory Position is displayed. Planetary positions are set through observations in observatories and corrected accordingly.

Birth chart cosmograph (horoscope), developed after computation of Lagna and setting of planets with the help of planetary position book (Panchang) Planetary positions of the day was taken and placed accordingly in respective Rashies. But moon was placed from Chandra rashi is mentioned before spasta surya (in panchang) above daily chart of planets. Now total interactions are apparent, i.e., up to 85, 30, 056 and more minute interactions calculated up to 1224440064.

These interactions refer to active, less active, inactive characters/may be genes of different natures (required, unrequired), some might have appeared in course of evolution. So, may be functioning, less functioning, etc. in nature analogous to present day study of genes.

The above calculations and horoscope setting are well described in books of astrology. Interested readers may go through for detailed study. Here only statistical/scientific portion is illustrated for conceptual preferences.

Belanter sarini for the computation of istkaal.

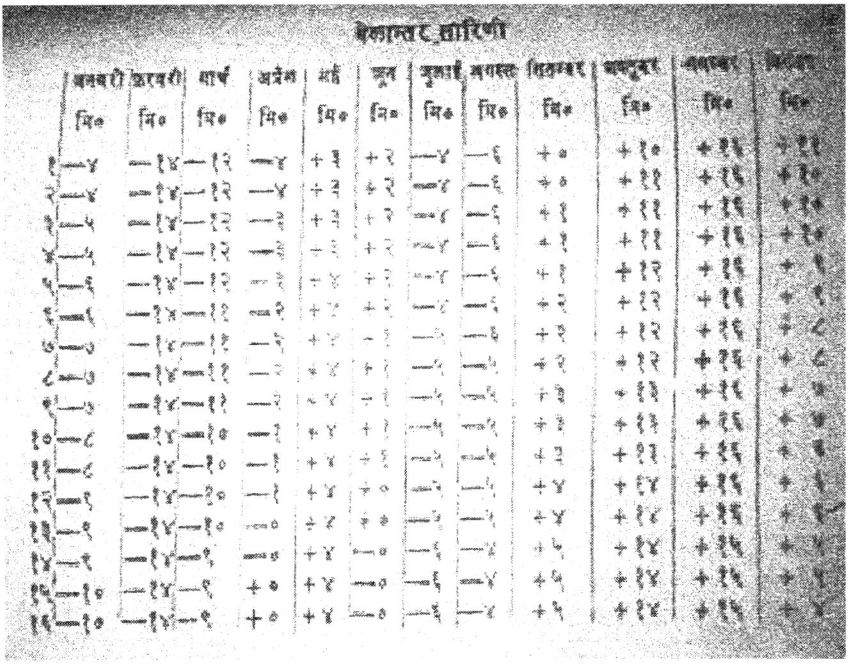

बेलान्तर सारिणी

	जनवरी	फरवरी	मार्च	अप्रैल	मई	जून	जुलाई	अगस्त	सितम्बर	अक्टूबर	नवम्बर	दिसम्बर
	मि०	मि०	मि०	मि०	मि०	मि०	मि०	मि०	मि०	मि०	मि०	मि०
१	−४	−१४	−१२	−४	+३	+२	−४	−६	+०	+१०	+१६	+११
२	−४	−१४	−१२	−४	+३	+२	−४	−६	+०	+११	+१६	+१०
३	−५	−१४	−१२	−३	+३	+२	−४	−६	+१	+११	+१६	+१०
४	−५	−१४	−१२	−३	+३	+२	−४	−६	+१	+११	+१६	+१०
५	−६	−१४	−१२	−३	+४	+२	−४	−६	+१	+१२	+१६	+९
६	−६	−१४	−११	−२	+४	+२	−५	−६	+२	+१२	+१६	+९
७	−७	−१४	−११	−२	+४	+१	−५	−६	+२	+१२	+१६	+८
८	−७	−१४	−११	−२	+४	+१	−५	−५	+२	+१२	+१६	+८
९	−७	−१४	−११	−२	+४	+१	−५	−५	+३	+१३	+१६	+७
१०	−८	−१४	−१०	−१	+४	+१	−५	−५	+३	+१३	+१६	+७
११	−८	−१४	−१०	−१	+४	+१	−५	−५	+३	+१३	+१६	+६
१२	−९	−१४	−१०	−१	+४	+०	−५	−५	+४	+१४	+१६	+६
१३	−९	−१४	−१०	−०	+४	+०	−५	−५	+४	+१४	+१६	+५
१४	−९	−१४	−९	−०	+४	−०	−६	−४	+५	+१४	+१६	+५
१५	−१०	−१४	−९	+०	+४	−०	−६	−४	+५	+१४	+१६	+४
१६	−१०	−१४	−९	+०	+४	−०	−६	−४	+५	+१४	+१६	+४

Pratham Lagna Table

Pratham Lagna sarini for determining Value of Calculated Lagna at particular degree in dand....pal.

For example...in the case mentioned above 4/13/6/11 is rashi degree of sun on 30th August 2016 (mentioned above). This means Sun is in Leo at 13 degree. Now following chart is used to know the Value of 13 degree Leo, which is 25 dand 30 pal as described just earlier. Now, ist kal was added i.e. 8 dand 50 pal, to become 34 dand 20 pal (derived earlier). This value 34-20 reflects as Libra between 0 to 1 in the table. So, Libra was selected as Lagna as mentioned above 7/00/00/00. This Pratham Lagna table is as following:

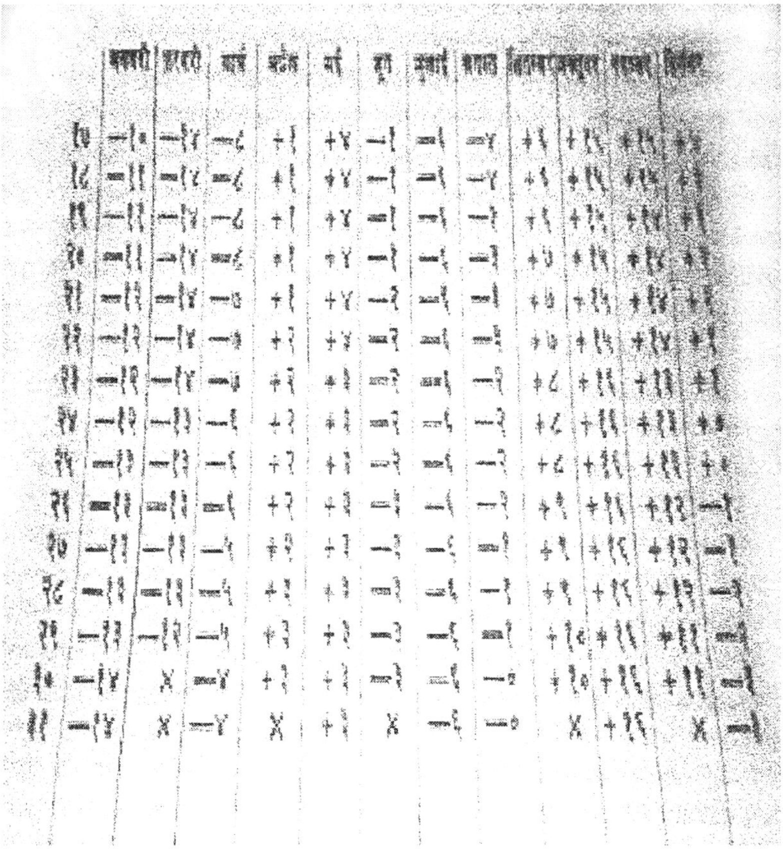

Note that the Lagna computed may be verified. 1st kaal is divided by 2. Now, constellation in which sun is present, that no. is added in the resultant. Again it is divided by 27. Rashi of residual no constellation should be Lagna.

This chart is again highlighted to digest 2 cycles

L- (I) and General I. This is slight enigmatic in substantiating interactions for both characters and disease analysis.

In the journey of preparation of Birth chart and then cosmograph, we have already discussed 3 points.

I. conversion of birth time into Istkaal, 3 -Examples with main variations and required tables

II. Lagna/first house/ L in cosmograph...computation with required tables.

For Rashi degree of Sun, moon Rashi and planetary positions, Panchang is required. Although essential tables already given.

III. Planetary positions in Birth chart, then COSMOGRAPH preparation.

Now...

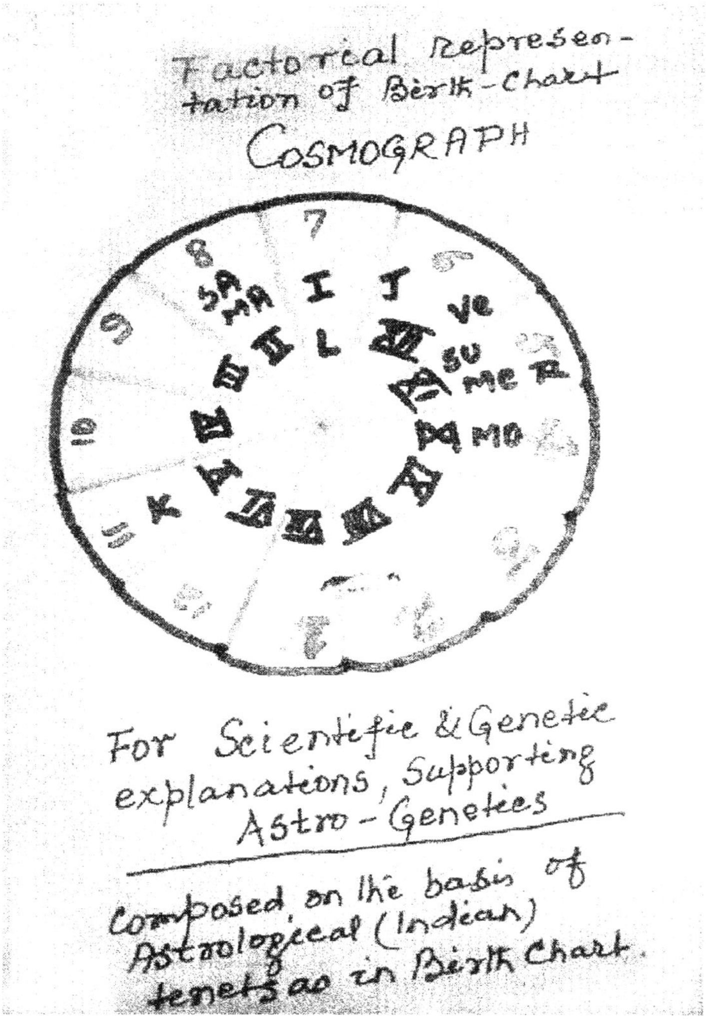

Factorial conversion of birth chart for scientific and genetic displays have been designated as cosmograph. This also follows every tenets of Indian Astrology. Cosmography presents 1-12 rashies. I to XII number of houses from Lagna L of COSMOGRAPH.

Planets indicated as factors for easy study of interactions like scientific studies. Statistics selected and are displayed here for recognising them only as scientific. However, they may be well studied from any Astrological Book. In this study main thing is to carve marvellous Genetic tenets in modern scenario from old studies as Astrology coined as interdisciplinary studies with Astrophysics, zoology, Life science, chemistry together. This paved an excellent breakthrough in Genetics as well in life Science for easy and cheap Genetic analysis and therapeutism of Human ailments. Long ago, cosmic rays were considered as any fantastic entity that might do only devastations and unknown miracles. But the modern science has brought it as estimated, weighed and used as "effect to cause" analysis. Ranges of effectiveness were measured and role also gradually studied. Now it is being observed as the only source of all events on the earth, whether it is regarding living system/non-living system and their interactions. The day is not far away when causes of Natural calamities will be declared as due to cosmic rays. Everybody, including scientists perceive the effects but feel defeated to accept its gigantic manifold actions. However, the study definitely brings science and religion nearer and apparent.

IV. Bhayat Bhabhog computation: Any individual takes birth during a constellation. The part of this constellation might have lapsed and a part remaining due to be undergone. The part already Lapsed before birth is called Bhayat. For this computation; period of constellation just before birth constellation was deducted from 60, and then 1st kaal/was added.

Example..... The constellation on 30th August 2016 was Pukhya, and just previous Constellation was Punervasu. The period of Punervasu was dand 13 pal....,13.

	Dand	–	pal
	60	–	00
(–)	13	–	13
	46		47
(+)	8		50
	55 d		37 p

= 55/37 –V_1 (as indicated in cosmograph)

V_2 Bhabhog total value of birth time constellation is calculated, then the previous constellation, (before birth constellation) is deducted from 60 dand and the resultant is added with the value of birth time constellation.

	Dand	Pal	
	60	00	
	13	13	
	46	47	
+	12	28	(Ist kaal)
	59	15	

So, V_2 = 59/15 (as mentioned in cosmograph)

Birth time constellation is Pukhya...12 d 28 p

Prev. constellation is. Punervasu... 13 d 13 p

So, Bhabhogah... 59 d. 15 p.

Value of bhabhog is greater than Bhayat.

If istkaal is added in above resultant (46-47). Bhayat is obtained, if birth time constellation value is added in upper resultant Bhabhog is obtained. It is in brief about bhayat – bhabhog computation (cited below).

46 - 47
8 - 50
55 - 37 = 55/37 = V_1

46 - 47
12 - 28 (birth time constell.)
59d 15p = 59/15 = V_2

Macro Cycle (Mahadasha)

Every Individual passes through 3 cycles of Some planets during his Life Span. In Binsottari system 120 years have been considered full life, which an individual very rarely attains. So, periods of some of the planets do not occurduring the full Life Span of most of the people. Every planet during revolution and rotation completes three cycles... macro, micro and sub micro (Mahadasa, Anterdasa, Pratyantar dasa). The chart indicates only Macro cycles of planets each during three constellations.

The following macro cycle chart displays Period of macro cycle of each planet during 3 Constellations. i.e. Sun 6yr Kritika; Uttarphalguni and Uttarakharha; Moon 10yrs within constellations...Rohini, Hast, Shravana Mars 7yrs within Mrigsira, Chitra, Ghanistha, Rahu 18yrs, Aardra, Swatee, Satvikha, Jupiter 16yrs within Punervasu, Vishakha, Purva bhardra, Saturn 19yrs within Pukhya, Anuradha, Uttarbhardra, Mercury 17yrs within Aaslesha, Jyestha, Rebti, Ketu 7yrs within Magha, Mool, Aaswani, Venus 20yrs within constellations. Purva phaguni, Purvakharh, Bharni.

Period of planets of an individual may be calculated first for Macro cycle with the help of previously calculated Bhayat Bhavog. The individual tipped as example, was born in Pukhya constellation from Macro period table given above, it is apparent that in constellation Pukhya, Saturn macrocycle remains effective. So, the period is 19yrs as already indicated in chart. Now, balance of the macro cycle is to be calculated at the time of birth.

So, bhukta aur bhogya dasa varsa (past and balance of the macro cycle) may be calculated, as following:

Example:

Bhayat - 16/39

(V1)

Bhabhog - 58/44

(V2)

Now, 16/39

60

960 + 39 = 999 V1

V2 58/44

x 60

3480 + 44

44

3524

Since Macrocycle according to constellation is of Saturn. The period of Saturn is 19years (Macrocycle).

Now, 999 V1

x 19 Macrocycle of sat.

= 599940

Again, $\frac{V_1}{V_2} = \frac{59940}{3524}$

= 1yr.

and residual 2470x12

= 29640 ÷ 3524

= 8 months

again residual = 1448 x 30 = 43440

43440 ÷ 3524 = 12 days

Thus 1yr - 8 months - 12 days

The period of Saturn is 19yrs - OM - 0 days

At the time of birth (-) 1yr - 8 m - 12 days

Sat. will continue 17 - 3 - 18 days 17yrs - 3 months - 18 days after birth and then subsequent periods will be added to calculate present period by also adding Date of Birth.

Note: Value of V1 (Bhayat) V2 (Bhabhog) are computed in very Scientific way with the aid of ultitude, Latitude. They are usually given in horoscopes either through computer or else.

Macrocycle: Each macrocycle consists of periods of every planet in noted series. Such as Macrocycle of Saturn consists of microcycles:

– Sat – Mercury – Ketu – Venus

– Sun – Moon – Mars – Rahu – Jupiter. (cf fig.)

Easy formula for calculation of micro period (Antardasha) is as following – Period of planets multiplied and divided by 10 to yield months again resultant is multiplied by 3 to get days of the micro period.

Ex : microperiod of Sun in Sun macroperiod:-

6 x 6 = 36 ÷ 10 = 3 months

Residual 6 x 3 = 18 So, 3 months 18 days is period of Sun microperiod in Sun Macroperiod though chart is available in every Astrological book. But shortcut calculation of sub microperiod (i.e., pratyantardasha) – In one microperiod, sub microperiods of each planet rotate.

Eg. Sun – Sun (Micro) consists of Sun – Moon – Mars – Rahu – Jup – Sat – Merc – Ketu – Venus. So, in Sun Macro period, Sun micro period: sub micro periods is calculated as all three periods multiplied and divided by 40.

Ex – 6 x 6 x 6 → 216 ÷ 40 = 5 days 24 ghati. This is the period of Sun submicro in Sun microperiod. This table is available rarely in good books.

Calculated chart each for micro and sub micro periods as examples: given below to calculate all others :-

Micro period in Sun macro period (6yrs)

Sun	Sun	3 m 18 days
Sun	Moon	6 months
Sun	Mars	4 m 6 days
Sun	Rahu	10 m 24 days
Sun	Jupiter	9 m 18 days
Sun	Saturn	11 m 12 days
Sun	Mercury	10 m 6 days

Sun	Ketu	4 months 6 days		
Sun	Venus	1 yr		

Sub micro period in Sun micro period.

Sun	5 d 24 ghati	Sun	Sun	Sun
Moon	9 d	Sun	Sun	Moon
Mars	6 days 18 gh.	Sun	Sun	Mars
Rahu	16 d 12 gh.	Sun	Sun	Rahu
Jupiter	14 d 24 gh.	Sun	Sun	Jupiter
Saturn	17 d 6 gh.	Sun	Sun	Saturn
Mercury	15 d 18 gh.	Sun	Sun	Mercury
Ketu	6 d 18 gh.	Sun	Sun	Ketu
Venus	18 d.	Sun	Sun	Venus

Gochar, Nawans are essentially required for knowing the effects of interactions or factors at particular time. This is known as predictions in Astrology.

In (panchang) present planetary positions are given. It should start from Rashi (where moon is situated in original). Positions of different planets may be read and effects presumed or prediction forecasted. Rashi acts as Lagna here. It is used for day to day effect analysis, known as Moon cosmograph. Nawans also helps in estimating gross effects at least related to fate ie aura resulting due to interactions with gene and cosmic rays further reflecting as electronic body; which in turn interacting with living and nonliving objects deciding activities in life. This is known as fate of the individual.

These types of cosmographs help/or essential for those requirements. Gochar or day to day movement of planets are easily procured from (Panchang) which has been corrected from observatory time to time. As indicated in Moon cosmograph Rashi stands as lagna. All planets counted from there and interactions studied. But Nawans horoscope is specially prepared. May be displayed later on.

Now,

Cosmograph....(cf fig.)

Factorial presentation of horoscope for scientific and GENETIC studies. Factorial transcriptions may be read as following:

MA....	Mars
SA....	Saturn
Me...	Mercury
K....	Ketu
V....	Venus
J....	Jupiter
R...	Rahu
SU....	Sun
Mo...	Moon

Two generations usually taken for study, where cosmographs have been prepared at sufficient time gapes on different places. Husband at one place, and after 6 – 10 yrs wife at another place. Then children after sufficient time gapes born at different places. In such family charts, traits (factors) have been found to be inherited (moving) from father to daughter, Mother to Son or father to Son.

In Example I

Analysis

(a) SU – Me → Inherited in

	(Father)	(Daughter)
	100°	100°
	5th general house	5th personal
(b) Mo (Father) →		Mo (Daughter)
	100°	100°
	5th general	5th personal
(c) Me – MA →		MA – Me
	Father	Daughter I personal
	VIIth personal	90°
	{MA – Me = Mother - 70°}	

(d) Factor	- R	- R	- R
	Father 10th	Mother	Daughter
	90d	9th	IXth personal
	(personal)	(personal 100d)	100d

Inherited

(e) K	Mother	Daughter
father (IV)	III	IIIrd
personal	personal	70°
90d	70d	

inherited

(f) SA Mother	Father	Daughter
XI personal	9th	3rd
70°	100d	70°

inherited

(g) J Mother	Father	Daughter
III (P)	VII (P)	VI (P)
70d	90°	70°

inherited

From above analysis it is apparent that factors moving (contributed) either from father to daughter or from mother to daughter. The movement (contribution) of factors from parent to offspring has been already established in modern life science as Inheritance in Genet ics (in live Samples). But in cosmograph this movement of factors appears peculiar and very much similar to live forms. Howcver, in Live forms this is happening in genetic materials (genes) but in cosmograph factors are merely cosmic rays and interactions of planets. So, they behave as genetic materials/genes and exhibiting a process like inheritance of factors from parent to offspring, thus bewildering and worth pondering over. Several such examples of Inheritance of cosmic factors are encountered, more or less in every family chart.

Example II

Mother	Father		Daughter
	Sun (L)	→	Sun (L)
	1st p		1st p
	II (me)		
	III (v)		
	IV (MA)		
	V (J)		
	VI SA	→	All houses
	VII		contributed
	VIII – J		similar
	IX – MA		
	X – V		
	XI – Me		
	XII – Mo		
J VII (90d) →	J (IX) P (100d)	inherited	–J (IX) P 100d
Mo XI (70)	inherited		VIII 70

Mo X P → Mo IV P

90° (father) 90° (daughter)

From Example II, it is evident that daughter Shares much factors from father. But factor J is contributed to Son becoming dominant in both parents. Factor Mo is contributed to Son from Mother.

Example III

	Father	Mother	D1	D2
(a)	MA (L)	MA (L)	MA (L)	MA (L)
	90d	90d	90d	90°

(b) J (VII) personal	J VIII P	J (VIII) P	J XI P
90d	70d	70d	70d
(c) R II P	R I (L)	R IX P	R IV P
70d	90°	100	90
(d) Me X	Me IV	Me IV	Me VI
90°	90	90	70d

When factor is dominant in both parents, it is inherited in all daughters. Factor (J) being recessive in mother is inherited as recessive in both daughters. Factor R being dominant in the mother is inherited in both daughters. It appears sex linked. Dominant - Me- in both parents is inherited in one of the daughters.

Example IV

Father	**Mother**	**D1**
K (L)	K (X)	K (L) (Inherited in same house as father)
90d	90	90d
	Inherited	

Father	Mother	Daughter
R (VII) P	R (IV) P	R (VII)
90d	90d	90d
J (VIII)	J (L)	J
70d	90°	XI 70°
	Inherited	

SA (IInd)	SA	SA II
70d	IV	70°
	90°	
	Inherited	

K is apparently dominant in both parents and transmitted dominant in daughter at peculiar L position like father. The character of daughter very much resembles father. Factor R is also inherited in daughter since it is dominant in both parents. SA factor, being recessive in father is inherited in daughter. Factor J is also inherited from father.

Example V

Single prominent factor inherited:

Father	Son I	Son II	Son III	Son IV	Son V
J (VII)	J (IV)	J (IV)	J (10th)	J (II)	J (IV)

Father has dominant (J) factor which is inherited in all 4 Sons, placed well in service except one having weak (J) factor. Father was doctor and one of the son was IAS, others doctor, Engineer, Professor. So, there is marvellous display, found in this family, confirming the inheritance of (J) factor. This factor definitely enhigh position of the individual in life (during 16yrs of its Macroperiod or even Dominant in Cosmograph) .

From the analysis of family charts it may be concluded that whole interactive surfaces of cosmograph appears to act as genome activating tools (energy) which stimulates human genome right from formation of Embryo to birth and there after developments and functions throughout life of the individual. The examples are not limited to 4/5. It may be traced in every family where statistical horoscopes/cosmographs are available. As the birth time and place are essential for its computation.

Thus the assertion partly may be promulgated as cosmic rays are causes of genetic inheritance. Since genetic phenomenon have been displayed during the process of inheritance so complete assertion is thus recalled as –

"Cosmic rays are causes of genetic inheritance and all genetic phenomenon through interacted energy support."

Natural phenomenon are consequences of interactions of Matter and energy only. So, Genetic material is matter and Cosmic rays are energy source to manifest all developmental and creative/ devastating phenomenon (events of nature) may be in correlation with inanimates, to exhibit magnificent activities and scenario of nature.

Moreover, there are evidences of genetic phenomenon during inheritance and traits development, as coupling-repulsion, Linkage, crossing over, dominance; recessive, loudly speak cosmic rays interference in genetic phenomenon and inheritance of characters. Thus, there is no doubt that cosmic energy is involved in genetic plays through energy supplementation to genes and also inherited in dominant or recessive forms.

Nature arranges everything in very simple design. So, genes developed in living system and powerful energy created. Consequently all activities going on uninterrupted and interactions continuously develop new characters for evolution. Eventually similar phenotypes of parent/offspring i.e., father resembles with daughter or son, and mother resembles with daughter or son are usually viewed in our daily life. When cosmographs are plotted of the similar phenotypic pairs, astonishingly both of the pairs show at least 70% similarities. Mother is very much similar phenotypically with eldest daughter and the younger is somewhat similar to her father. These things appear very often and people take it for granted as Allah or God has blessed. Not that, there is mere genetics in it. If they have records of their birth time and place, genetics may be exhibited after going through pure statistics and cosmographic laws in the tune of old Astrological principle after scientific scrutiny. The pair found nearby that mother and eldest daughter were 70% similar was analysed through cosmographs. In both L/ was same, meaning there by all houses similar. Some important interactions were also similar L in both having J x M x R - interactions in 1st (L). In 8th there was J - Mo interaction. Phenotypically both are of white complexion. 1st to 12th houses (rays) were similar except some planets situation different.

These are the veiled truths of nature but confined under cover of religion. It is thus claimed that in every family there are phenomenon but unanalysed because in that respect science is still juvenile, as exploration are not targeted to reveal everything. Expedition goes on Moon, Mars etc at high costs but only for wealth and politics never thought of its cosmic contributions and interactions those are helping to thrive humanity here. So, in this pace modern science can't unveil these crude realities/ of nature and its ultimate, activities atleast regarding one solar system. What to talk of one galaxy and then total galaxies, to comment upon the existence of God because he may be anywhere or nowhere. How can one emphatically claim? This is also unfair drama of nature in respect of science that as it goes up to touch the peak, pretty before delusion occurs and everything starts again from zero. And these riddles remain unsolved - forever. So, the discovery of genetics is marvellous contribution of knowledge (of modern science) with old sciences and ultimately to Adhyatma. After all, it may be asserted that Genetic body and cosmic rays are only mysteries of Nature to answer all riddles. In the illustration of K factor inheritance one more very interesting phenomenon is observed. In the regulation of L (1st) house both in W - H by same SU factor speaks something special. This SU factor in both (L) might have acted as a trait of attraction during marriage negotiation unknowingly through the natural process of selection. In this connection, it is to be emphasised that unless strong force of attraction becomes active either by situation or by virtue of period in the absence of negative forces wedding is not fixed or materialized. In the same example IV inheritance of combination factor (SU - Me) is displayed. This factor is powerful in both W - H. In husband it is present in IVth when in wife it is in VIIth, both 90d reflecting the character in potent way.

In the same, J factor inheritance (example Vth), one more interesting genetic phenomenon is found (G I plate). He was himself a doctor, having 5 Sons but no daughter. Wife was Simply a house wife, only literate. After analysis of cosmographs very good and interesting inheritance of factor (J) observed (discussed ealier) for all-round development of father and Sons. But mother did not contribute. So, this might be the case of sex-linked inheritance.

Sex linked inheritance: (Example V plate)

Analysis: Father was borne on 22.5.1935. His L (lagna) was 12th Rashi Presenting typical –J– factor inheritance in his family ie. among his sons and himself.

D.B.	**Lagna**	**J factor. Pow. (L)**
Father, 22.5.1935	12th Rashi	7th 90d
1st Son. 20.4.1957	3rd Rashi	4th. 90d
2nd Son. 11.8.1959	5th Rashi	4th. 90d
3rd Son. 27.9.1961	4th Rashi	7th. 90d
4th Son. 22.7.1963	11th Rashi	2nd. 80d
5th Son. 27.2.1967	1st Rashi	4th. 90d

It is to be reminded again that houses 1st, 4th, 7th and 10th are centres (kendra), assigned with power 90d, 5th and 9th are triangles, assigned full power 100. In the above chart, father and four sons were having – J– factor in centres (kendras) except only one son who was having J in 2nd house but in the house of J itself. Really he was a bit inferior among brothers, (not much). 2nd house is bad having 80d power, when in this case power is referred as 80, simply because J is in its own house, so power increased (swagrihi).

Dominant J of father inherited well in dominant positions more or less in all. It is apparent from powers of this – J – factor assigned and effects also phenotypically studied in real life. Study tools were prepared long back at different periods (Horoscopes and Cosmographs). This appears the case of sex-linked inheritance, because mother recessive and no daughter borne. So, astonishing inheritance was displayed by his father to his sons. It is beyond doubt, and can't be suspected as chance.

In the example IV:-

SU is present in Xth house and Me in IXth. Showing genetic phenomenon of crossing over within MO, MA, Me, SA, SU factors. However, Me and SU both are in powerful houses ie in 90d and 100d, justifying parental Rich attribute in those factors. Inheritance of factor... R.... in all three cases may be analysed. It is present in powerful house (90d in power) reflecting as potent Character in all. IN Father it is present in VIIth House,

Daughter also inherited in the same House VIIth. While it is situated in IVth in mother. Power remains the same in all.

Inheritance of Factor.... J .... It is weak in both Father and Daughter, but strong in mother (90d). In father it is 60, but in daughter it is 70d. In Inheritance of Factor SA:Daughter again inherited, this SA factor from IInd of Father to its IInd (70d) while it is in IVth in the mother. 90d. Consequently, Eldest daughter has inherited most of the characters from Father. So, now we can see that all cosmic rays are ultimate causes of Genetic phenomenon through Energy support; as engine to the rail bogies through generations. Cosmic rays are definite scientific tool, not fantasy, its power, angle activities are now well known and well within the records of science. It is recalled because whenever cosmic rays is talked many people become flushed into fantasy then everything becomes unscientific and imaginary. So, it should not be confused, it is definite scientific entity.

Sun (cf. fig) along with other stars, Constellations, Rashies and gallectic nebulac perform all activities in living and non-living systems through powerful tool cosmic rays.

They have been found to be Causes of all human diseases and Genetic phenomenon. Uptil now we have discussed role of cosmic rays in handling genetic phenomenon but its role in causing human diseases will be proved in chapter II, i.e., "Therapy of Human" diseases. So, the above slogan appears to be completed (.......).

Graphical presentations of the interactions of Cosmic rays with body physiology and Genetic materials are contained in cosmograph. This is the picture of an individual develops instantaneously at the time of first oozing out of head from Mother's womb i.e., birth (relationship of this individual with celestial bodies ie planets, Rashies, and constellations at particular angles and degrees). For details exhaustive discussions and displays have been made but yet to be completed. Essential photographs and diagrams cited (all original based on self-study and experiments for 25 -30 yrs). In brief, it is the depiction of the mechanism which Nature adopts for controlling living and nonliving systems and their interrelationships. So, it is vast but truth of life. Much part of it resembles dharmshastra as kan-kan me byapt (omni present & all powerful). Latest part of science specially Genetics advances towards some similar kinds of results.

In any family member when viewed, categorically it appears showing genetic phenomenon, in the resemblances of phenotypes from either father or Mother. In some cases behaviour, speech, working style of Son or Daughter resembles with father or Mother. These are genetic tenets inherited in Son - daughter from father - mother. Everybody accepts these genetic behaviours but how these tenets are reflected in family cosmographs showing inheritance of cosmological factors/ interactions from Mother/Father to son or daughter. This is not limited in one or two cases. So, that it may be called "by chances"? This is present in many cases rather in each family cosmographs chart the genetic phenomenon are observed. The inheritance of potent and specific factors exhibit marvellous inheritance, (when Cosmograph is simply a chart of planetary displays). How does this happen? So this conclusion and Slogan is 100% correct that cosmic rays are causes of genetic phenomenon (as discussed previously).

Discussion

Since the beginning of civilisation Indian Astrology has been well accepted by common people. So, it became a part of our culture. Moreover, this field has contributed many things to modern science regarding planetary positions, process of observation through observatories, knowledge of eclipses and their exact time calculation. Mathematics/statistics and there uses in planetary calculations for agricultural and different calamity uses, asteroid effects, digits from 1 to 9 and zero invented by Indians only and then whole world adopted. In Veda i.e., in Rigveda "chakra" word was coined which indicated Rashi chakra of 12 rashies. Thus a rich contribution made in modern Astronomy, mathematics and statistics besided others (cultures and rituals).

However, this branch is still treated as religious/ superstitious and occult, Simply due to ignorance of facts or lack of knowledge regarding Natural attributes (cosmic rays, planets, stars, Rashis, constellation). That appeared as main reason for this science to remain undeveloped (or becoming up to date). Is this not appearing like hypocrisy that we all (modern science also) accept, Sun, Moon and other planets and at the same time different behaviours of different rays. Radiation biology is well developed. We are now capable to explore detailed biochemical behaviours of living systems. But unfortunately we ignore the most vital aspects which are quite pertinent in controlling human behaviours which may easily be analysed when a geneticist goes through Astrological details. The effect of moon is well under modern science i.e., lunatic (a type of madness).

This is the adverse effect of full - Moon. Oceans show marvellous peaks of tides. Witnessing all these how can a developing science ignore vital natural phenomenon or powers just to impose supremacy or in the name of classical and modern. Again this is recalled. We are all surrounded by high energy particles but modern science ignores, when they already know the effects of energy but don't consider its effect in their theories and experiments. How? Human behaviours are controlled by rays/planets perhaps through Gene regulation. This gap is actually due to faculty-wise specialization in Modern science. For the sake of convenience in study, we have drawn a line in modern science that this is Botany & else - But really speaking can a botanist alone describe every aspect of a plant? Plant has interactions with so many phenomenon as root with nematodes, bacteria, etc. Root is geotropic and Stem - Apo geotropic so gravitational force (physics), plant undergoes interactions with numerous chemicals of soil (so chemistry). Thus, a scholar having been awarded Nobel prize in all subjects of the faculty can only explain a plant in true sense. So, wisdom of convenience we have adopted to ignore vital aspects. Thus again recalled we are all surrounded by cosmic rays (1 to million bevestron) and modern science ignores it and advocates theories and Laws. Is it not hypocrisy? How our functions, biochemistry and Genetics remain uneffected under such powerful energy influx. It is high time for this matter to be pondered. Thus, a line though appears revolutionary, has tried to trace for the sake of reality and unadulterated knowledge.

Basic Tenets of Genetics

Genetics is an important branch of life science dealing with the inheritance of the living systems. It is also a science of prophacy a bit. On the basis of the characters of the male and female parents we predict the characters of offsprings of different generations. Laws of dominance and segregation (Mendelian Laws) are evident in different pedigree displays. Even law of independent assortment has practical applications.

Cosmic rays as potent environmental factor:- Minute observations disclose the possibilities that the human race appears a combination of :-

1. Heredity - 50%
2. Cosmic rays - 40%
3. Other environmental factor - 10%

ratio may vary slightly under scientific design. However, this needs confirmation.

Interplay of heredity and Environment is well documented where vital aspect Cosmic rays has been completely ignored. The process of inheritance is completed at the time of syngamy. Consequently, the zygote undergoes development and differentiation under direct control of environment where most vital component is cosmic rays (either through utilization of living green biomass and through astrological results in different generations described earlier).

Basic Tenets of Astrology

There are five branches of Astrology out of them first two are -

1. Theory (sidhant) and Hora (Ahoratri), having least error and appear scientific in modern sense.
2. Birth time of the individual is the exact time of emerging of head of the baby and it is standarised considering Altitude and Latitude of particular place. Thus local time converted in standard time and first house Lagna in particular Rashi is computed. That becomes individuals

Ist house (L) besides general 1st (Mars) for every individual.

3. The constellation (Nakshatra) of that period gives the idea of Macrocycle of particular planet. After calculating balance of the constellation through Bhayat - Bhabhog calculation. Period passed in mother's womb is bhayat and bhabhog is the period of constellation baby will enjoy after birth. From this juncture life clock starts in a cyclic order of planets Macrocycle - Microcycle - short cycle - then another Macrocycle of another planet. How much scientific is every stage may be judged impartially through these incidences of life science?
4. Further, each rashi denotes particular organ of the body which is controlled by brain. There are 12 rashis and similar pattern is found in Human physiology when interestingly 12 pairs of cranial nerves control different organs of the body from brain (fig).
5. The first pair of cranial nerve supplies to the same organ as that indicated by Ist Rashi, and so on so forth (from endocrinology +).
6. Subsequently brain also controls more organs of the body through 27 important spinal segments. In Astrology also there is concept that 27 constellations control minute organs. Here also the pattern of supply to the organs is more or less similar (cf Endocrinology)
7. So, Bramhand (cosmos) has been considered as Brain which controls Human body through rashis and constellations. As the human brain controls the organs through 12 pair cranial nerves and 27 important spinal segments.
8. Planetary positions in horoscope of the individual determine angles, power, and interactions.
9. In Astrology, there is concept of calculation of the periods of planets in Bingsottari and Astottari system. Bingsottari system considered 120yrs. standard age of individuals while Astotthery based on 108yrs age. Bingsottari system has been considered more appropriate so adopted here.

10. A particular Macrocycle of a particular planet switches on only certain characters good/bad under effect for action, likewise, switched on genes.

11. For evaluating over all effects cosmographs are prepared variously - Moon cosmographs starts from the rashi of the moon & then other planets placed. Situation of moon in the rashi is also called as rashi of the individual and considered as the Ist house and accordingly others. This gives idea of present positions of planets and used to know day today interactions keeping in mind the regular ones. This moon prominent horoscope is also called as Gocher.

    The planet revolves taking whole period to cover complete path, say as 18 - Rahu, 19 - Sat, 6 - Sun, 10 - moon etc. These are macrocycles. During this course of journey, each planet rotates at its axis where interactions of the rays become different with all planets one after another called sub-microcycle. So, in the journey of a planet in Macrocycle each step is micro-cycle. Every planet has power to control through reflections of cosmic rays. Every planet reflects that is called (sees) 3, 5, 7 houses except Jupiter Saturn and Mars. Jupiter in addition reflects on 9th, Saturn 10 and Mars 4th and 8th.

12. Confluence of planets are highly specific forming a part of genome study. They are dealt in books as 2 planet interaction, 3 planet interactions, 4 planet interactions, etc. may be a part of genetic study and gene (traits) characterization. As also owner of one house is characterized as situated in different houses, i.e., owner of L in 2nd, and forming part of Genome study. So, different types of characterizations are found in Astrology. As, whole 360° comprises of good, bad, less active, fully active, inactive genes/traits formed through different sorts of interactions exhibited.

13. Macrocycle of a planet in its Long duration changes the nature and behaviours of individual as according to its characteristics. This confirms genetic activity and genes into action during the period.

14. In the normal situation also most powerful planet (or interaction) turns the individual accordingly (as according to the characters of planet/interaction).
15. Rays are effective in three layers of the individual. In the broad spectrum it interacts with human physiology, but higher intensity effects internal organs and at highest intensity for longer duration causes genetic changes/ diseases and rarely mutation.

Planets Full Action (Reflecting Process)

Microcycle is the small cycle which also changes interaction with same sets of planets. As Sun Macrocycle - goes through all moon, Mars, etc. microperiods and in Sun - Moon again all planets.

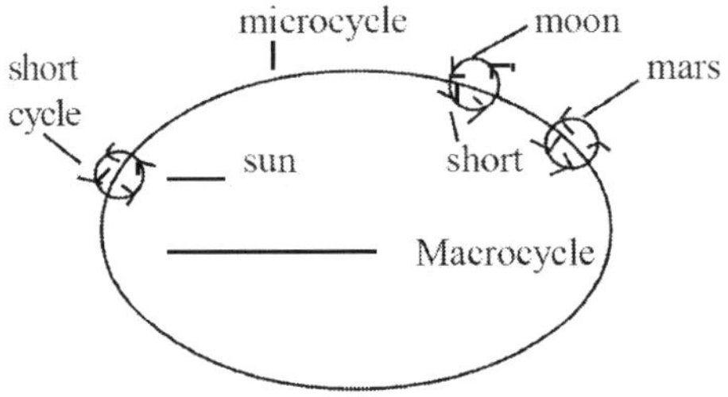

Macrocycle, microcycle and sub microcycle complete revolution on the path is Macrocycle. One step forward in which interactions change is called microcycle. Planets rotate during revolution. So, each rotation during first step, i.e., Microcycle is the sub microcycle where also interaction changes in the same sets of planets in brief.

As SUN macrocycle goes through all planets ie. moon - Mars, Rahu etc microperiod and in SUN - moon, microperiod all period rotates ie. SUN - moon - mars, SUN - moon - Rahu, SUN - moon - Jupiter, SUN - moon - Saturn so on. Two planets remain same, third interaction changes name as sub microcycle (Pratyantar dasa) 2nd type of cosmograph is called Lagna cosmographs in which Lagna computed earlier labelled as 1st besides general 1st (Mars). If Venus is L, then Venus is 1st, Mercury 2nd, Moon 3rd so on for individual. But according to general cycle for all Mars - 1, Venus - 2, Mercury - 3, Moon - 4th so on. Both cycles interactions and Gochar, are considered while planetary effects and effects of cosmic rays are taken into account for disease diagnosis or others.

Planets are good and bad - As according to primary characters - Jupiter, Venus, Moon, Mercury are considered good - but due to bad angles become bad in action and power. The owners of 3, 6, 2nd 12th, 8th houses are bad. There is immunity power of owner of L. If owner of L is owner of 6th/8th, that becomes neutralized to some extent and owner of L remain good. Angles - 1st, L, 4th, 5th, 9th, 10th, 11th are considered good except planets of 11th are considered bad to be regulated during disease therapy - A.G.

Three layered characteristics of planets also support the genome study and interactions with cosmic rays.

Three layered effects of cosmic rays on body by each planet:(the above and present characterizations are analogous to gene/genome of cosmograph).

Jupiter

- Foot, thigh, liver, digestion, blood pressure, nerve (part) - organs.
- Good behaviour, generosity, beauty, liver, devotion, power, intellect, wisdom
- Temples - worshipper, Minister, Judge, university, contribution, sympathy.

Mars

- Outer brain, skull, nose, chick, contagious disease, wound, operation, pain, blood infection.
- Courage, firmness, anger, self-confidence, quarrelling tendency, Influence.
- Army, Doctor, Professor, Engineer, chemist, barber, carpenter, blacksmith, Manson,
  Technician, Games and Game article agents.

Moon

- Bowel, digestion, breast, uterus, vagina, eyes, nerve, genitellia.

- Sensation, internal desire, impatient, domestic attitude, imagination alertness, Gainning attitude.
- White colour, ship, fish, water, liquid, nurse, servant (maid), diet, silver, Blue.

Mercury

- Brain, nervous system, tongue, speech, hand, artistic
- understanding, memory, logistics, fine art
- School, college, science, scientist, courageous works, art publications, editor, writer, publisher, postmaster, businessman, learned fellows, yellow colour, effect of mercury.

Sun

- Heart, blood circulation, eyes, small arteries, teeth, ears
- Power, dignity, love, generous, ambitions, self-confidence, self -control, thought, imagination, balance, benevolence
- King, minister, head of dacoits, invention, geologist

Venus

- Neck, kidney, shape, fairness, hair, beauty, body, conducting organ, penis.
- Affection, knowledge of beauty, start, love, cleanliness, testing ability, working capacity
- Beautiful, ornaments, pleasing, dance, decoration, artistic materials, love, motherly, attainment

Saturn

- Bones, Lower teeth, Large intestine, muscles.
- Agriculturist, dak peon, grazier, miser, police, officer, fasting, saint Hilly ears, Rock, desert, cave, old, dilapidated place, grave yard, Field.
- Deep knowledge, Advices, Freedom, Element, Memory, Working ability, Temperness, firmness, admand, Gravity, character, Wisdom, Alertness, Genuine, Leader, patience.

Rahu (Dragon Head)

- Sleeplessness, stomach, Mental trouble, madness.

Ketu (Dragons tail)

- Skin disease, mental trouble, loss of appetite. Whether these detailed descriptions do not tell upon the analogies with genome study? It becomes more appropriate when experiments satisfy and results approve genomatic behaviour. More scientific effort in this direction will pave quick and better results with improvements. The experiments will have far-riching effects on Humanity and Environment very quickly. So, it is claimed that cosmic Rays are causes of all human diseases and Genetic phenomenon.

Cosmic rays are causes of genetic phenomenon is more or/ less satisfied (cf chapt. I). The next part will be proved in chapter II.

Development and differentiation is directly correlated with protein synthesis in Living system. There is currently a great deal of interest in unravelling the molecular mechanism involved in cell differention or the acquisition of functional states by cells already differentiated, and hormone actions are an aspect of this problem. Formation required the synthesis of specific enzyme and other proteins those are coded as sequence nucleotides in the chromosomal DNA and the expression of this information could be coded by regulatory agents acting at the level of transcription or translation. Generally assumed that the mitotic process in the various cells of multicellular organism will possess the same genes but it is obvious that they do not have the same assemblages of enzymes and proteins. The mere presence of a gene (through heredity) does not allow that it will be transcribed into RNA and translated into polypeptide chain on the poly ribosomes. The cell must contain regulatory devices that determine which genes are activated and at what time along the axis of development (cf. endocrinology). The presence of "repressor" and "derepressor" that turns on and "turn off" particular genes has been postulated. It is promoter, gene which determines the site for the synthesis of m - RNA since it conveys direction to RNA polymerase about structural genes. But who determines promotor? Answer is factors. The promotors are once decided just at the time of birth or syngamy. So, at a particular time particular gene is activated under the influence of particular cosmic rays (under any cycle). This ultimately causes protein synthesis.

Protein is electrically charged particle exhibiting attraction and repulsion controlling Social biochemical and all sorts of behavioural activities of human race. Electric charge consequently exhibit attraction $\rightleftharpoons$ repulsion which is quite evident. A person under the influence of strong Saturn shows quite repulsion toward a person passing through strong Rahu (Dh) pertaining to the fact, determination of promoter at one time selects biochemically the fate of the Individual.

Turner, Writes in his endocrinology that outpouring of hormone (e.g., ACTH) is influenced by intense light, sudden temperature changes, sound and strange environment leading to an effect on adenohypophysis and thus several diseases. But he ignored most vital force cosmic rays or strange environment may be interpreted like this. Such inclusions at the threshold of Astrological evidences is not at all vivid rather a matter of statistics. It exhibits timely and potent effects.

As discussed that revolution of planet on long path completes Macroperiod of a planet and its rotation completes micro and submicrocycle. Individuals clock starts at the time of oozing out of head from womb of mother at the moment a period of constellation also lapses and rest (bhabhog) starts eg for factor MA (Mars) three yrs may pass (bhayat) in womb and 4yrs may be enjoyed (already discussed). So, a macrocycle of a planet remains under command of a protein and its nature changes with the change of Macroperiod through corresponding change in promotor. However, in micro period and sub micro period, out pouring of the same hormone may be inhibited/stimulated based on the nature of planet in relation to the individual. The intensity of the effect varies man to man in accordance with the nature of planet displayed in one's cosmograph (remain effective).

At the onset of a particular cycle the gene under influence may be characterized and thus evaluated with least standard error. But intensities of effects are based on experience.

Thus, the nature of protein synthesized (hormone synthesized) during Dragon's head cycle must be different from that of Jupiter cycle. So, the individual becomes completely changed, sometimes rash behaviour developed during (DH) turns into good/nature. (often observed/ AG) Even accidents are caused due to planned synthesis and hormonal control which may be well analysed through Astrology/Genome study.

Sudden minor changes perhaps upto sub microcycle (pratyantar dasha) is caused due to effect of rays on hypothalamus, which controls the behaviour through enzyme adrenaline. Thus any change in the period leads to an effect on hormone either on the synthesis or control circumstances which an individual is bound to obey.

Hither to, Astrogenetic tenets converge on a point that key of human body is protein (hormone, enzyme, etc.) and change in its nature and control is a predetermined phenomenon which can be well estimated through scientific tools. So, any research in this direction will benefit modern Genetic Engineering, Medical science, Astrology and ultimately to human race. Eventually this is veiled mystery of nature which has not yet received any attention.

The latest technique of cats scanning is based on histopathological basis, so modern science is still behind reality (causes). Talk of gene cloning for causes analysis of cancer undoubtedly a step in the right direction. However, Astrology displays evident idea of gene of an individual through cosmograph and its approximate time of development. There are 12 (houses) x12 (Rashies) and 27 constellations. So,12 x 12 x 27 x 27 x 9 x 9, possible types of combinations appear which may be the types of genes. However, functional status of a gene and its regulations is a matter of calculation whether through changed promoter or through inhibition - Induction process at exact time.

Thus beyond doubt, Astrological concept is based on more advanced technological concept (interdisciplinary ideas suggest) which modern science may or may not attain in the present speed and manners. What is indeed wrong if modern developed technology undertake Astrogenctics i.e., interdisciplinary concept for speedy progress of human race with high mental development and power therapy.

What is FATE? A Scientific (Astrogenetic) Concept

Although 9th house from L (Lagna) is allotted as fate (according to Astrology) but true interpretation is the over all situation and confluence of planets, their angles formed, and Switch on planet (bhabhog of planet) at the time of birth. Regarding fate, this is general belief that we can't correct it. It is true upto some extent, but we rectify it in the sense that we add strength to weak rays or we trap it more with the aid of R.A.D. and simultaneously bad rays are eliminated from adding evil design of effects also through R.A.Ds. The angles formed from some rays (planets) remain bad, meaning there by such rays even if essential are not received. So, R A Ds, make them available to some extent. For this usually precious stones are used. This is enough to a great solace at that juncture for suffering Individuals.

Besides, acquaintances of the future incidences, this knowledge, alert us and give courage to face the situation or adopt safety measures as far as practicable. This is what we call fate and our true "karma" based on all possible efforts by self wit and wisdom. Sometimes we feel that for a particular work, even after making sincere efforts, don't get success whereas in the favourable time tough work too is automatically solved. The knowledge of future incidences may also make several facilities available to the individual which may save his time and energy. Sometimes it is also observed that few good students suffer from malefic aspect of bad planets and face disturbances regarding finance, health etc consequently their career become ruined. Such disturbances may be overcome to some extent by the uses of RADs. However, there are some wrong aspects of its implementations. That some idle students/ person without making effort try to solve problems by using RADs only. They always suffer and can't achieve anything out of it. This is immature design of thinking that without labour everything is achieved only through RADs. These applications avert hurdles only. One should be very clear that death can't be averted similarly definite incidences. Simply the amplitude is reduced that much comes to a great relief in life. So, correct and timely analysis and use of Astrogenetic fact and RADs help a lot.

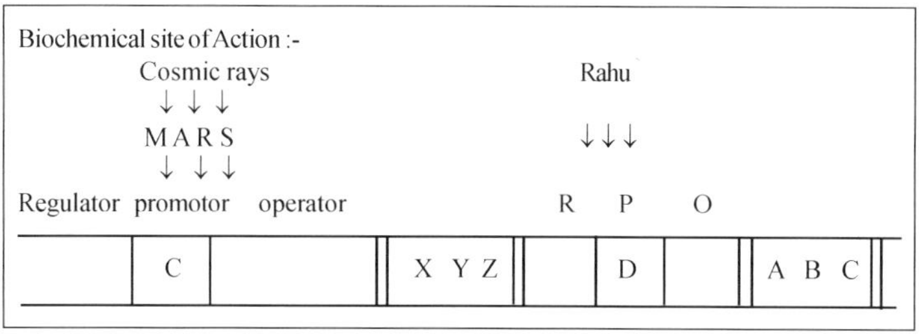

Nature of m-RNA

2

Health and Wellness (Therapeutism)

Exogenic Regulation of Human Diseases (Cosmic Rays are Causes of all Human Diseases and Genetic Phenomenon)

Society became fully engrossed in rendering services to ailing mankind young boys and girls of all disciplines approached to learn this therapy. Among them Science graduate/Arts graduates and also some medical students (of MBBS course) came to learn on regular basis. The society was voluntary so there was no fee for teaching, besides there was no privacy in sharing mechanism and formula like old vaidyas and Gurukuls. For simple maturity sake Graduation of any discipline formed the basis of study except Medical students. So, a syllabus of Interdisciplinary course was developed–

Relevant portion of different subjects selected to swallow the theme –

SYLLABUS (3yrs +1 Practical)

(a) Chemistry – Primary - Element, compound organic, inorganic, minerals, periodic table, solid, liquid, gasses, crystallization, matter, energy, there inter-conversion, chemical reactions.

Advanced - chemical composition of stones, effect of rays on specific chemical compound and an attempt to explore effect of rays on precious stones (saphire, emrald, Diamond) and mechanism of action in respect of the human body - So far practicable, effect of some rays on liquid and gases. Brief history of chemical evolution and origin of earth (A.G. concept).

(b) Physics: Characteristics of rays, colour effect sun rays (vibgyor), gravitational force, x-xay, B ray, r rays, ultra violet rays. Reflection, refraction; light, electricity, Sun, stars, galactic nebulae, earthquake, electromagnetic energy.

(c) Astrophysics - Cosmic rays – nature, origin power and mechanism of Action, Electromagnetic rays, role of cosmic rays in the evolution of earth, Aura formation, Kirlion photography and behaviour, ozone layer, ultraviolet rays.

(d) Zoology – Detailed study of Human physiology. Control system through cranial nerves and spinal segments.

Human pathology – primary knowledge

Human diseases – primary knowledge – organs involved, protein matabolism, operon concept, Glimpse study of the animal kingdom and evolution. Tool study B.P., thermometor, weight meter. Primary ideas of human pathology, different pathological tests.

(e) Cytology and Genetics – Human cell and its inclusions. Conceptual study of Genetics - Mendelism, Interaction of genes, Linkage, Coupling - Repulsion, Genetic Mapping crossing over, recombinants, Genetic Engineering, Human Genome, gene, genetic diseases, oncho gene and Cancer, operon concept.

Diseases

Special - primary knowledge of cancer, AIDS, muscular distrophy, Schizophrenia, Spandollosis, Mental retardation, Parkinson, Alzhimer, Titanus, hydrophobia, deaf, dumb etc. chronic, sicle cell anemia. Seasonal disease, chronic diseases, Rhumatoid Arthritis, Rhumatism.

(f) Astrology - Systematic and conceptual study in tune with genetic studies and cosmograph studies :-

Rashis, Nakshatra (constellations) charans and alphabets group of Rashi- Nature, friends (Similar) and foe (dissimilar) owners of stars of rashies of planet characteristics. Stars of constellation of planet characteristics. Organogenesis according to rashis in correlation with cranial nerves; organogenesis according to Nakshatra (constellation) in tune with spinal segments (important) cosmograph computation - conversion of local time into standard, ultitude - Latitude, Istkaal, Belantar Bhayatam - bhabhog (past and residue of Nakshatra) First (individuals) House ie, Lagan (L) calculation setting in mind the cycles 12 houses (from L - individual) and II nd from general 1st (mesh). Situation of character from both angles confirmed. Special combination and their results - Like deaf, dumb, eye defects, aristocracy, service boldness, richness, poverty, business magnet, landlord, gains from marriage place, issue, marriage widower, gains from father, Longivity, Foreign tour, These traits are gaining support from both.

Nine planets with moon (Gochar), bingsottari calculations for periods may be from charts, or from formula. Thus, macroperiod, microperiod and submicroperiod calculation take place. Then mount reading verifies positions and interactions E (elevated), D (depressed), ef (effected), bad angle, etc. more study with interest.

Life Science: Simple knowledge of Human Pathology and diseases (organs involved), tool study (B. P. Temp. weight. primary ideas of ultra sound, primary idea of different scan, MRI test, Pet scan, terminal diseases (organs involved and basis).

(f) Practical: Tool operation, patient dealing, diary maintenance, doze weighing and doze preparation. Heartbeat, B.P., temperature, etc.

(g) Implemetations:

a. Disease cure/disease cycle

b. Characterization of genes, no. of genes for different characters

c. Location of genes in Cosmograph and human genome

d. Inheritance of genes (view from cosmograph).

e. Character building (from theif, daeoit drunkard, drug edict to normal)

f. Genetic engineering

g. Quality Improvement, circumstance manoeuvring.

h. Lengthening of life span (effort may be made).

i. Real and Scientific Gyan of every event of Nature, relationships (wife, father, son issues neighbour etc.)

j. Relationships with nonliving objects.

k. Natural calamities.

l. Friends & Foe.

Note: Some of the topics of few subjects remain unfinished due to lack of research facilities.

Ethical Committee

ETHICAL COMMITTEE

Certified that Astrogenetic Society is rendering free Services to ailing, mankind. This therapy is most relevant , Scientific and highly effective. It covers a wide range of diseases in a very simple way. It is protective as well as curative and potentially imparting a hope for rejected and dejected patients. This Astropathy is also suggestive of daily health care, saving from drug abuse, by utilizing nonconventional source of energy. Risk and side effects are no problems as dozes are reversible.

The present project will definitely be proved as a boon for humanity in general and ailing mankind in particular. It will also systematise our present knowledge.

1. Dr. Surendra Pd. Singh M.B.B.S; M.S — Regn No. 15525 Surendra Prasad Singh, Physician, 5.9.96

2. Mohan Krishna Verma, Retd. Distt Judge — Mohan Krishna Verma 23.9.96, Retd Distt Judge, Ex President, Distt Consumer Forum, Darbhanga

3. Ugresh kumar Jha — Advocate, High Court, Patna, Registration No. - 354/95

4. [illegible] — Social Worker (Mukhia)

24/9/96

Area Electricity Board, Darbhanga

Dr. Shravan Kr.

डा० श्रवण कुमार

Dr. Shrawan Kumar

M. B. B. S. (D. M. C. H.)

CIVIL ASSISTANT SURGEON

Member, Faculty of I. I. M., Calcutta

Ex-M.O., B. C. C. L., Dhanbad

Specialist : Kalazar, T. B.

PHYSICIAN & SURGEON

क्लिनिक एवं मिलने का समय :

दरभंगा, गंगवारा :

सुबह—७ बजे से ११ बजे दिन तक

एवं आकस्मिक बुलाने पर

तारसराय :

दोपहर—३ बजे से ६ बजे संध्या तक

आकस्मिक किसी भी समय

Name : Date—

Age/Sex BP/Pulse

Certified that

Astrotherapy is conducted at Astrogenetic Society, Kathalbari, Darbhanga by Dr. C.H.Jha. since 1982. It has been recognised by I.C.M.R. New Delhi, Science Council Japan and other high scientific institutes of the world e.g. A.I.I.M.S., New Delhi. It remarkably renders great services to humanity scientifically to atleast rejected and dejected patients. This therapy has very high percent of recovery than any other exhisting pathies. However, Astrogentic regulation covers psychological, medical and paramedical /Genetic diseases, besides preventives, I am personally much impressed with the activities of the society.

Thus I render my services voluntarily to this society which has distined to serve ailing mass of the world selflessly.

Civil Assistant Surgeon

Bihar Health Services

For therapy organogenesis is must. A great similarity was found in rashies and cranial nerves. There are 12 rashis and in human system there are 12 pairs of cranial nerves. The attached (controlled) organs are similar in both. Ist p. cranial nerves is attached to the organ same as Ist rashi denotes and so on so forth. There are 27 constellations and there arc 27 important spinal segments miraculously both refer to the same organs. So, both rashis and Nakchatra complete, organogenesis, this helps a lot in therapy exhibits pattern of control system (Human physiology) through body physiology.

Human Physiology

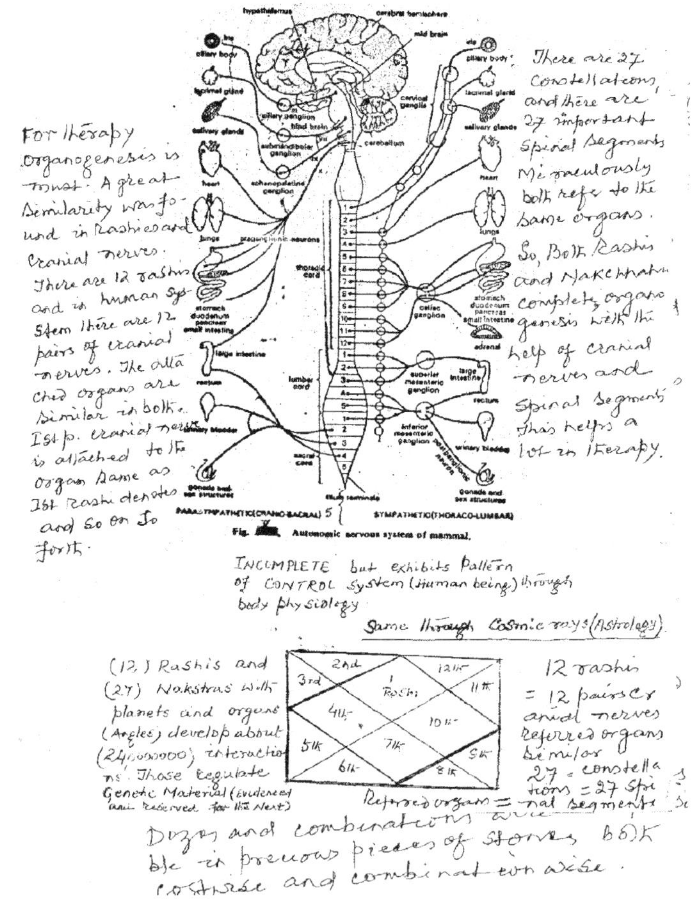

Some Thoughts Regarding Cosmic Rays (Astrology)

(12) Rashis and (27) Nakshatra with planet and organs (Angle) develop about (8503056) interactions. Those regulate Genetic Material (Evidenced and received for the next)

Referred organ = 12 rashis = 12 pairs Cranial nerves referred organs similar.

27 = constellation = 27 spinal segments. (important).

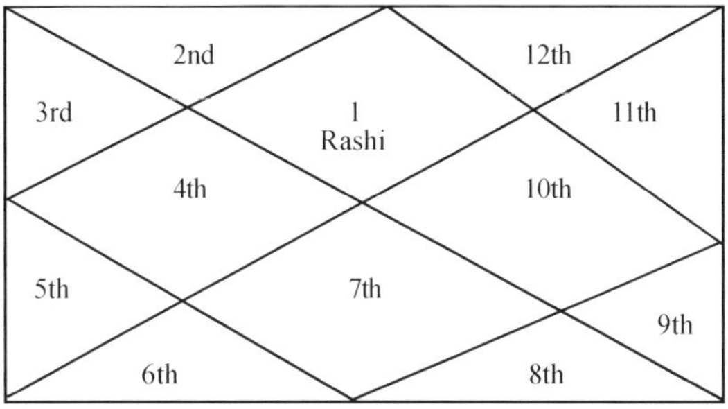

Since the time immemorial persons using stones specially affluent section of the society they use for two purposes, either to elevate personalities or to change their fate. This practice very much prevailed in the society and then people of every walk of life and strata such as Doctors, Engineers, academician's even students started wearing stone pieces in gold or silver or copper rings. Some people in 20th century started also using for health. However not categorically for particular disease regulation, cure and dozes pattern. In late 20th century when people felt slight effect of stone pieces in diseases like cheering effects not exactly curing, then we started a novel practice first of all in family itself. There was a clear cut idea in mind that rays are mixed in nature coming after several interactions so single stone can't cure any disease, organ is involved, causual organisms may be there and body metabolism (physiology) may be depleted. So, exploration turned towards combination dozes on the pattern of telegraphic message. For giving effect to one point, it was started capturing all effects there giving special effect to the main. So, therapy started concentrating upon dozes and combinations which was not possible with precious pieces of stones both cost-wise and combination wise. Consequently, we switched over to semiprecious stones. Now problem of endorsing combination dozes were indeed became serious. First of all semiprecious stones were tested and found that some of their good quality responded well. Then accidently came across some beads wholed for the use of garlands. They were all semi-precious stones. There we started new era of combination dozes, keeping in mind stitching (encountered). By the way fishing fibres made of nylon came to sight in the market. Prior, thread (normal DNC) and other were tried but not succeeded because patients had to keep the dozes 24hrs tied on Arms and Neck. The nylon white fibres could gain all-round support. Threads of silver, gold were also tried by lady patients but they were proved uncomfortable and impracticable.

Five years elapsed in the researches over these things and to coming over to exact dozes of diseases. Astonishing effects were found with these combination dozes in minor diseases. So, gradually number of patients and family practice increased. After all, dozes pattern in tune of diseases discovered which was really unique in the world and easy to combat diseases. Increase of the dozes started covering diseases corresponding number, combination and weight. This type of doze - disease dependent time bound treatment was never discovered/ recorded through these stones (Semiprecious/precious). 1990 onward full fledged treatment started of almost all diseases (except tetanus, hydrophobia and AIDS). Even rejected and dejected patients started coming from every part of the country and treated successfully. Still this type of treatment neither found in the world nor recorded anywhere till date. When full swing treatment started ICMR, New Delhi guidelines for human patients handling were followed thus Ethical committee was formed (copy attached). So, gradually dozes evolved in accordance with necessities and diseases. But scientific paraphernalia, above tool and technique were thoroughly maintained. However society remained voluntary with support of free technicians and helpers. Still this unique combination of dozes bearing person comes to sight he is identified as a patient of Astrogenetic society. Society always worked on the principle of service to man is service to God.

Tools: Tools Required in the Therapy are Simple but Scientific

1. Thermometer - Hicks/digital
2. B.P. Instrument - old pattern with mercury
3. 10 X glass for observing stones (RADs)
4. Small Balance (shown in fig.)
5. Weight box in Ratti and cent (Here used but it may be converted in grams, milligrams). It should be kept in mind that weight must be very correct.

   10 cent = 11 ratti (in this way the dozes were converted).
6. Threads - Nylon threads (used in preparing fishing net).
7. Electric wires - thin wire for fixing RAD beads in thread.
8. Weight measuring machine (for patients)
9. Tray for RADs
10. Small towel
11. RAD Box - Semipreceous stones (RADs - ray altering devices) Some stock of RADs be maintained because they were not easily available in good quality. For general use shops be directed to keep them in good quality. They are limited in number:
    a. Coral (beads of Italian with whole)
    b. Garnet (beads wholed)
    c. Topaz (beads wholed)
    d. Pearl (beads wholed)
    e. Hessonite (beads wholed) - Gomed - beads wholed
    f. Lepis lejule (Lajbrat) - beads wholed
    g. Tiger's eye - bcads wholed
    h. Cat's eye (beads wholed + piece of ring)
    i. American diomond - Ring pieces.

Requirements of few ring pieces of preceous RADs become, essential for Genetic and terminal patients. Dozes were prescribed according to patients requirement & body weight.

1. Yellow/white Saphire - usually required
2. Emrald - pieces for Rings/beads wholed
3. Manik beads rarely/cheap manik/Star Ruby.

Next essential factor is to maintain Patients Diary even if it is a case of own family. It will not out of place to mention that there was not at all expenditure of drug in my family. Every ailment was immediately regulated through dozes even if it is 12 in night. Some of my friends and relatives/patients were also using (after a training) after 1990. Some of my friends after taking proper training started sub- centres for ailing mankind. Sri Nripendra Nath Jha was running centre at Bokaro Jharkhand India. I myself treated ailing Mankind at different centres in India on the demand of that region e.g., Samastipur Bihar, Patna Bihar, New Delhi, Mumbai, Maharashtra on the demand of patients. Everywhere camp was voluntary except place managements by organiser.

Patients Diary

Name and address of the patients were noted at the RT top corner. Below that Mount study was given if cosmograph of patient was available it was recorded and period, interactions observed. Etiology, body temperature was taken by thermometer. It is highly remarkable that temp. 97 was only considered normal, Lower and upper temperatures were taken seriously unlike medical Science (which considers 98 normal). The view implied that many small ailments are within the range of temp 96-97-98, which Medical Science ignores or tells psychology, but patients feel and suffer. Similar attention was paid in B.P. readings. 120/80 - 110/70 - 100/60, were very seriously scrutinized and dozes depend on these types. Temp was taken (as normal checkup) after taking breakfast, never empty stomach.It corresponds heartbeat. Heartbeat, and pulse were also taken seriously In case of any abnormality advance tests were suggested. For Heart, Chronic patients haemoglobin test was suggested specially to ladies. They were usually suggested calcium, and iron in case of discrepancy during further treatment to keep blood level maintained. Weighing- machine (shown in photograph). Primary readings were taken very seriously giving full time which becomes foundation of diagnosis. Attention was also paid to see medical treatment history, if returned after medical treatments of institutes.

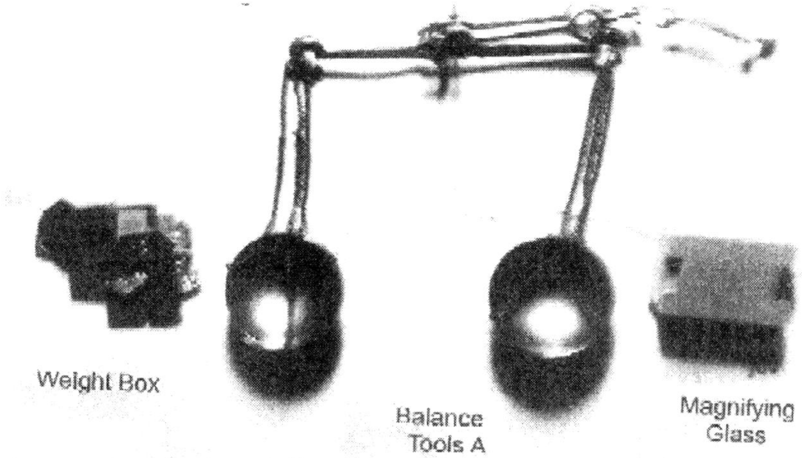

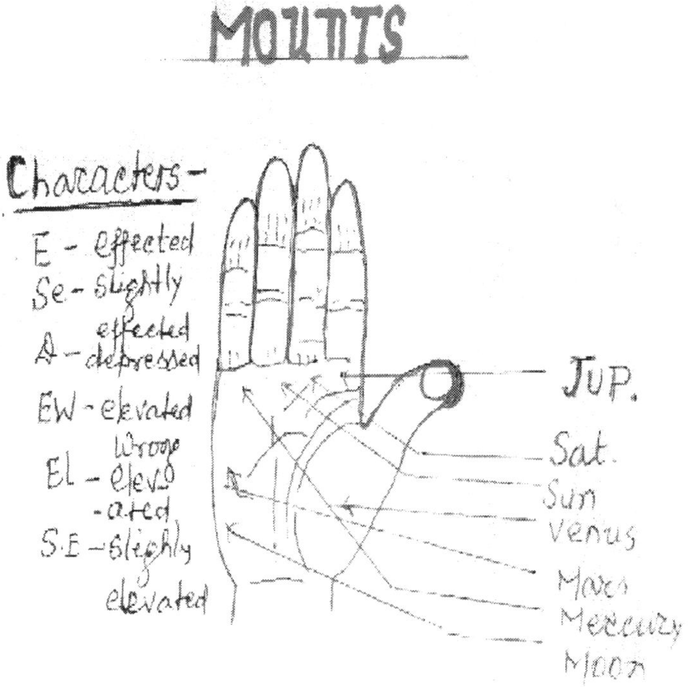

When stones cure ailments how do medicines affect and cure? Medicines also by virtue of its chemical, molecular power reduce (like umbrella) the bad effects of devastating rays from the organs, till a definite period of time if within such period disease vanishes, it is considered that medicine acted positively and cured the diseasc. But actual mechanism becomes apparent when the effective time of Medicine dozes becomes over and period of devastating rays remains as such, the cured disease again appears. This is apparent in the case of chronic diseases. Drugs in those cases endorsed for the whole life, e.g., hyper tension, blood sugar, thyroid, uric acid, heart ailments, etc. meaning there by the chemicals of drugs suppresses effects of diseases by forming a curtain from bad rays but power of good planets become weak i.e. immune power deteriorates. Here it is recalled that immune power is the gross effects of good planets in the individuals. So, effects of good planets decreases and bad planets increases this condition is congenial (suitable) for diseases to appear. Thus here the period of bad effects was the period of disease. As soon as bad effects of bad planets become over the disease vanishes in degrees. So according to durations of bad effects of planets, the characteristics of diseases may be of 4 types:

(1) Seasonal

(2) Chronic

(3) Heriditory

(4) Mutant (rare)

Seasonal and chronic diseases sometimes appear as if due to some worms, virus, bacteria etc. S0, some of them are called contagious. However, they are secondary causes, primary causes are Rays only. This has been substantiated after long discussion with scientists of higher institutes. In this pretext, it was concluded that bad rays when bring some metabolic changes in the body or organ that changed or effected protein becomes suitable for the growth and Life cycle of that pathogens. As soon as period of bad planets becomes over it induces or restricts Metabolism of bad medium/proteins affecting badly the population of pathogen which diminishes gradually to insignificant or nil as the case may be. This brings relief in disease. So, causes are bad rays only, even in these types of diseases. Hence depending upon the duration of bad effects of planets the disease may be called as seasonal and chronic. If bad effects persist for days, it induces seasonal disease, but if the disease is due to some bad interaction of planets or due to specific situation of the planet which is non-reversible, may cause chronic disease. Sometimes medicines cure such ailments but as soon as medicine is stopped. Ailments again crop up. In that case Astrogenetic therapy is also continued for long or whole life. If ailments are due to bad macrocycle of any planet, say for 19, 18 or 20 Years, these ailments also called chronic. They become cured after bad macrocycle changes. But sometimes, ailments become nonrecoverable due to production of bad protein/metabolites during the bad long period of macro cycle. So, these are seasonal and chronic diseases.

As mentioned earlier, patients diary was maintained as... Name, Age, Ref. No, Date... Adress... Place... B. P. .... Temp. ..... Wt. ... Pulse...., Sex, Homoglobin...Jup...E/SE/Wrong/D/SD, Sun...Do...Merc. ... Do... Venus.... Do... Mars... Do, Rahu... Do, Moon... Do... Saturn...Do.

Any Ailment - Present - available pathological/radiological records, record of surgery if any are recorded. Mounts are the receptors of rays so characteristics develop and studied to analyse power angle and vital interactions. There are interactions which cause diseases. Hence, Cosmic rays are causes of human diseases so palm reading were recorded in patients etiology diary as mentioned above.

Mounts marked well elaborated (cf fig.). Then ailments of the patients were recorded with important pathological findings. Any notable fact regarding previous treatment was noted. On the right side cosmograph and periods were noted. Below all, dozes were written and still. Below that, spaces were left for writing reports each after 48 hours (for serious pateints).

Palm reading as given earlier in brief was just to refer complete process. This procedure was adopted for both if cosmograph was available with the patient or not since it was compared with hand readings. In case it was not available then hand reading formed the basis. Attempts were explored to procure cosmography provided date of birth, time of birth and place of birth were recorded by parents or else at the time of birth. It is customary to note in many Indian families. Moreover, it is now mandatory to note at the time of birth in hospitals also for birth certificate. Anyway, if it is available then patients are told to go for its preparation either manually or through computer. However, emphasis was given on manual. Simply because of the planetary details available in computers of remote south. If it was normal then computer computation was also suggested. The time taken in these formalities was shorted out in the way as treatment started and patient was suggested to make it available. But if the patient appeared of grievous nature treatment stopped till formation of cosmograph, (if all records were available). In case of non-availabilities of records, palm and pathological records were formed the basis. In any case results of first doze revealed all truth veiled in the cosmography. So, palm readings were sufficient for the treatment. Cosmograph helps in suggesting future course of disease.

For palm reading only mounts were considered, not lines as palmists. Normal mount was considered at central rt place, neither excess elevated, nor depressed. Any effect also becomes visible. So, remarks were noted as below, index finger. Mount of Jupiter for that. E (elevated) or D (Depressed), effected (apparent depression on the mount), if wrong in angle, mount becomes slightly displaced written as wrong. Below middle finger is Saturn. Everything was noted for this planet also. Again below little finger everything was noted for Mercury. Just lowest edge from Mercury is Moon and below thumb is Venus. Everything was noted for them also.

The characteristics of planets and groups of Stars, i.e., Rashies and constellations were found similar. So, planets were considered owners of Rashies and constellations in Astrogenetic concepts. The characters described at many places may be the locations of genes. Now, human genome analysis reveals such characters, these facts will be more elucidated after depicting genetic phenomenon in the pedigree cosmological charts. Abbreviations used in page 23 - Ma... Mars..., Mo... Moon..., ... Me... Mercury..., Ra.... Rahu...., .. Ke... Ketu..., or Kentuke, Ve.... Venus, Jup.... Jupiter, SAT... Saturn., ... Su.... SUN...,

Bowel disorder, low range diabetes, back pain. Arthritis sciatica, etc. are cured along with developing immunity or as a result of immunity development.

Genetic Diseases/Hereditary

I was called by Genetic Department, AIIMS, New Delhi for a discussion on Astrogenetics, well after summoning Liaison Committee meeting by Science Council, Japan on this topic and consequently declaring the subject quite new and told to Peruse this research with their good wishes. I, along with some technicians visited AIIMS, New Delhi in January 1994 (letter enclosed). In the first session of our talk, we were asked to participate in a conference organized by Foreign Nationals on a Genetic disease Muscular dystrophy. I was acquainted with the disease and its genetic aspect. Moreover the impact of stone combinations and dozes were worked out on few patients with Positive effects on one of them. The disease is due to loss of genes so not functional. Actually genes are not lost, they are well at their places but not functioning perhaps due to loss of energy/non availability of energy for formal skin manifestations, due to time and place of birth. The supplementation of energy to the site of those gene action, gave positive effects. So, there is a list in Medical Science for genetic diseases, but they are hardly doing anything positive. They are engaging patients, doing some primary things, like exercises as in mental retardation cases. Just time pass, so, they became ready for collaboration on the instances of ICMR, New Delhi, which assured financial support under project. So besides medical range Astrogenetics added some more diseases to the list. It was found that genetic diseases were also caused due to abnormalities in genetic behaviours and functions. Astrogenetically, it is considered cosmographic errors, i.e., wrong position of the planet not supported by others and effect of long macro cycle of quite adverse planet. Astrogenetic finding added that diseases which either parent or both suffer may be transferred to new born' also at the time of birth. Many genetic diseases fall under this category also. So, genetic diseases are due to parents or fore father in both maternal or paternal sides. They contribute as direct or hidden traits reflected in offsprings, sometimes under genetic distributions and inheritances, but in some others transcripted by chance, during genetic transmissions/inheritance. Sometimes assessed roughly called the disease due to direct activity of genes, in whatever circumstances it may be Since, there is involvement of genes, medical science is doing least in cure process. So, Astrogenetic therapy has done and will have to do a lot. But percent cure in some diseases observed where Medical Science was silent. Diseases like muscular Dystrophy, mental retardation, Asthama, T.B., Arthritis, madness, sterility, haemophilia, diabetes, B.P., cancer of some types alkeptoneuria, sickle cell anaemia, Parkinson, etc. are genetic diseases may be due to bad interaction or adverse period for long duration or may be due to dismal immunity of the patient. These diseases are most tough diseases. Medical Science is silent, managing some any way. No recovery. We are also doing in percentage 60/70. Maximum 80% is dependent upon the patience of the patients in long therapy. Genetic engineering and biotechnological researches on synthesis and transfer technologies may pave substantial ways.

We have already discussed that rays are good and bad, due to good and bad planets, good and bad angles i.e. rays. Coming from stars of bad planet characteristics or coming from bad angles are called bad rays. So, when these bad rays regulate any organ of individual at particular time for longer period, diseases develop in that organ or simple physiology is affected. (If organ is not in question). Usually such condition develops when they can't cure them, giving only soothing effects. The cure is only possible after switching off the bad rays. This is when the diseases are caused due to bad period of the rays. But when the diseases are caused due to interactions, drugs are suggested for the whole life and endorsed in such cases. RAD dozes are also endorsed for the whole life. So, diseases are caused during bad period of rays and become over when rays switch off V.R. interaction causes allergy. So it is due to interaction effect. In such cases bad effects persist always to some extent but aggravate in the period of those planets. There medicines have suppressive effects only. Medicines or stone dozes suggested for whole life. Mechanism of action of drug and RADs are similar i.e. through chemical in both as Drugs are assimilated so they effect temporarily whereas stone dozes are exogenous and never assimilated so their effects continue. Since, drugs don't have perennial and full effects so endorsed regularly and doctors use other resources like operations, physiotherapy, diets, exercise etc. So, cosmic rays are primary cause of diseases. Drugs bring temporary relief through chemical actions because they are assimilated. But RAD dozes are applied exogenously give continuous effects through ray altering power due to chemicals without being assimilated. Besides therapeutic power, RAD dozes have effects on immunity development and body strength make up mechanisms too.

Mutant Diseases

Irradiations are sources of mutations, i.e., having mutagenic effects but it is not so common on earth surface, however, it is not quite uncommon. Some experiments which carried out on cyanobacteria flora by Jafar Nizam and Puspa Shrivastava, confirmed that few cases of mutations occurred, when the test organisms were exposed in cosmic irrediations in space through ballon flight though very little contributions are on record. Likewise people on earth crust also face such effects but very rarely. In this regard Astrogenetic finding is that the planets with long macrocycle such as Saturn... 19yrs. Dragons head (Rahu) 18yrs. Venus 20yrs Mercury, 17yrs. Mutation is inducing, however, the most effective is Saturn (which during 19yrs if very adverse) produces abnormal mutant protein in the body sometimes whole body becomes twisted high pressure or high degree of diabetes, osteoporosis of high degree. Osteosarcoma and, other cancers, etc. are resultant diseases. Likewise patients Nature is also altered during that long span. The adverse Rahu induces madness, dacoit like behaviour, drunkard terrorist, heinous, inhuman deeds, paralysis, Parkinson, etc. Venus within 20 yrs causes sexual disability, (if very adverse). High degree of allergy, may develop, sexually transmitted diseases up to AIIDs with the help of most adverse Rahu. Mercury being adverse may cause (within 17yrs). Tuberculosis, heart problem, diabetes, spinal problem, may be twisting, brain troubles. The main thing observed that during most adverse position and nature of those planets effects induced remain un altered.

Jupiter is also a good planet but due to its adverse position and nature serious Lever diseases, blood diseases, some contribution to B.P., blood sugar are also not ruled out. Anaemic conditions also manifested under very adverse conditions of Jupiter. Astrogenetic therapy may bring some soothing effects in the ailments, percentage cure were observed, since it was difficult to break mutant proteins So produced within such a long span of times. However, if a patient would have come for regulation in the beginning they might have been saved to a great extent by Developing immune system of the body through a particular doze. So, dozes have also preventive measures up to some extent in a range of 5 - 45 units. When regulation takes place for immunity development they become free from regular cough, fever, B.P. pains, and headache. Combination dozes 1st and 2nd may be of same combination but of increased wt. and number in Right Arm (cf. photograph). This is quite initial doze, Only a few no. of patients to whom it suits. The characteristics wise their SA, R, factor are weak. So, 1st doze may consist of Italian drilled corals 5 ratties + Garnet beads 5 only... on Rt. Arm. This first doze may be started with 3 + 3 in case of an infant up to 3 years of age. This doze if suits may be increased up to 10 + 10 and sometimes a third stone is required, i.e., Topaj in proportion of 3 ratties in 5 unit and 6 ratties in 10 ratties doze. It is rarely required in 3 ratties doze. Dozes corresponding diseases will be described later. However in adult 10 ratti doze may be used in Rt and left both.

Tools Used in the Therapy

1. B.P. Machine (mercury type)
2. Balance
3. Wt. Box (cent and Ratti)
4. Patients wt. Meter (not shown in photo)

Small balance: It is easily used in weighting stones, more or less accurately but not in comparison with electronic or chemical balance. Uses of those balances were not practically possible. Here, cent and Ratties are used, because these weights rhythmically correspond to dozes as have been practiced and set for dozes. Thus, this simple and less costly balance was selected for use. This was also considered that patients may purchase balance who were suggested to use dozes for the whole life.

10 cents = 11 ratties, if ratties not available cents may be used with this formula.

Next dozes - may be of 10 units, 12 units, 15 units and 20 units. Upto 10 units right Arm is used. Specifically 5 units doze is given on Rt arm only (adult). In children 3, 5, 7 unit dozes are endorsed on rt arm. However, in adult 10 unit doze may be used in rt arm only or rt and left both, 2 stones (coral and garnet - 10 units + 10 beads) on rt arm and 2 (Lepis + Hassonite - 5 rattie each) on the left arm. In case of more doze requirement garnet alone is increased 1 or 2, 3, 4, 5 beads on rt arm. For such increase heart beat and temperature are monitored to reflect 97° or 72 approx. beat. Sometimes no. of garnet may be twice the wt. (ratties) of coral. If increase of single garnet does not reflect suitable then 12 unit doze is endorsed in both arms. In 10 unit left - Rt doze, third stone may be required, i.e., Topaz (6 ratti in 10 unit). Sometimes 4th stone is also required in the case when this is the final doze i.e. Pearl (2 - 3 ratti). Requirements are always judged through meters (B.P., beat and temprature) Moreover, patients gesture and soothingness also count. Multiplied use of garnet be cautiously tried since it gives pressure on heart directly.

If there is requirement it should be definitely increased otherwise after trial of 1, 2 increase it should be taken back. There is also a process for taking back of any bead. In that case full doze be put off. First dozes of the rt hand is put off then that of Left. Again after 10 minutes left doze is endorsed and after 2 minutes rt doze (after withdrawing extra garnet). First of all, body temperature of a patient is taken in Morning hour after taking breakfast, never empty bowel. That gives idea of a real reflection of the body. In the astrogenetic therapy normal body temperature means strictly 97° any deflection is marked seriously, whereas in medical science 98° is normal. That is the reason why minor ailments of patients remain untreated in Medical science. Patient suffers but doctors say psychological.

Sometimes in this range, major ailments of patients such as pains, allergic manifestations e.t.c are hidden there. In left arm also 3 - 4 stones are added such as:

(1) Lepis by wt

(2) Hessonite by wt.

(3) Tiger eye - it is by count. upto 15 unit 2 beads of medium size and 20 unit doze 3 beads. Sometimes 4th stone is added by weight i.e. cats eye.

After 10 unit, 12 unit doze, added rarely only in rt. usually in both arms. Then 15 unit doze is given in both arms, next is 20 unit doze, if neck dozes are required it is added not in full doze rather only in 20 + 20 numbers (onex and cryst.).

The left-right dozes are suitable to the patients with strong Saturn. Detailed use of dozes in corresponding diseases will be elaborated later on in both arms (+) or (-) neck dozes. Wearing of doze starts from left arm, then after 3 - 5 minutes Rt. arm dozes then in neck. Dozes may be set anywhere on the arm but in the same plain on both the arms. The putting off of the dozes starts from Neck, then rt arm then left arm. Dozes corresponding diseases will be discussed later with cases of real patients.

The above mentioned doze is altogether different from low B.P. patients (100/70, 90/60). Many dozes of the above combination are suitable for rheumatic, nerve troubles and some gastritis patients. If rt - left combinations does not suit due to body characteristic then left- neck dozes endorsed (low B. P. dozes).

This left - neck doze is endorsed on left arm first and then on neck. On neck only combination of two stones, (crystal + Onex), and in left arm three types of stones in combinations for all i.e. Lepis, hessonite and tiger eye. Further abrief presentation/ acquaintances of new combination of dozes, never witnessed or formed the basis of any therapy in the world, has been illustrated. The range of dozes and their combination are discussed as primary exploration of dozes of this new therapy. The most wonderful aspect is the exogenous application of all dozes, nothing for intake but affecting more quick on the internal site of action. So, one doze may be used for several patients one by one provided it is not advised for the whole life or very long period. This is the beauty of this therapy rather most economic part. B.P. temperature and pulse are monitored to witness the effects of dozes.

Doze Pattern (A)

Since the dozes applied are totally new in the world, so primary acquaintances and some rule regulations described.

Now detailed description of all dozes (illustrated) is as follows:

Pattern A – Rt – Left + Neck

1st Doze - is made up of 5 units in Rt hand only consisting of 5 ratti (coral) + 5 beads (Garnet).

Prior one special doze was also used of 3 units for infants and children below 5 yrs of age, consisted of 3 rattis (coral) + 3 beads garnet.

2nd Doze - has only 10 units rt i.e. 10 ratti (coral) + 10 beads (Garnet) only.

3rd Doze - has only 12 units rt i. e. 12 ratti (coral) and 12 beads of (Garnet). These patients usually have cold effect more on left side. Topaz may be added (5 R to 7.2 R)

4th Doze - Now, in balanced patients left + Rt requirement starts.

Left		**Rt**	
5 ratti + 5/5.5 + 1		10 ratti + 10 beads	
Lapis	Hassonite bead tiger	Coral	Garnet

This is not out of place to mention that A. G. doctor never gives any doze rather patients body requires, which is judged through meters and doze pattern. Doctors can't endorse even a single bead excess or wrong. A wrong doze immediately shows imbalance in tools (thermometer/ B. P.instrument). A doctor should simply feel requirement of the body that only he observes.

5th Doze - So, 5 ratti Lepis + 5/5.5 Hassonite + 1 bead tiger in left and 10 ratti (coral) + 10 beads Garnet in Rt and 5/6 ratti

Topaj may or may not be added in Rt and 7-2 ratti be tried in next 6th doze if required (in Rt.).

Left			**Rt**	
6 ratti +	6 +	1	12 ratti +	12 beads
Lapis	h	tiger	Coral	Garnet

6th Doze -

Left			**Rt**	
L +	H +	T	12 ratti +	12 Beads
6.0	6.5	1	Coral	Garnet
R	R	b		(extra big size)

7th Doze - Left + 15 unit rt

(L + H + T) 15 c + 15 G + Topaz

9R 11R 2 b ratti beads 7.5 ratti

Serious patients were directed to report anytime = (at inconveniences if beyond expectations or already told). General patients when reported, all checkups taken again before endorsing 2nd Doze. First Doze was given for 10 units, and second doze will be of 20 units. Minor ailments either persisted or cropped up were taken into consideration that ailments will subside in next doze. Decision of the next doze is taken considering all aspects rigidly. Since dozes if increased, becomes difficult to withdraw because it takes time to settle and it is a bit troublesome to patient. So, at this juncture, there are options. Patient may be given only 10 unit doze. Thereafter garnet is increased empirically. 2 beads or up to 15 or sometimes upt 17.... 18...20. This is increased usually after testing with thermometer. If garnet is increasing much, increase in size of the beads may be tried. In the transition of one to two dozes, one more point is to be considered seriously that for so many patients 12 or 15 unit dozes may be suitable. In that case, 10 unit doze appears suitable at first, but after 3 - 4 days if slight trouble crops up, then attention to be paid towards upper dozes. That's why patients are kept for 10 - 15 days. In This therapy there is provision of easy reversal of dozes with slight trouble and time. In case of confusion during therapy dozes/diagnosis, patient should be put on right track (on digisted dozes), then automatically body will demand right dozes through minor tools (temp. B. P. Pulse). So, it appears that this treatment is body treatment diseases automatically subside.

8th Doze

Left			**Rt**		
L +	H +	T	Coral +	Garnet +	Topaj
10	12	3	20	20	12
R	R	b	R	Beads + Extra garnet (if needed)	R

9th Doze (Variation may be due to body wt.)

Left			**+ Rt**	**+ Neck**	
Lapis +	H	+ Tiger	20 units	crystal +	onex
12	13/14	3	as above	20	20
R	R	b		b	b

10th Doze

Left	**+ Rt**	**+ Neck**	
10th dose	20 units	crystal +	onex
as above	as above	20	30
		b	b

11th Doze

Left			**+ Rt 30 units**			**+ Neck**	
L +	H	+ T	30 +	30	+18	Onex +	crys
18	21/22.5	3	coral	(Garnet)	Topaj	30	45
R	R	b	R	b	R	b	b

12th Doze

Left			**+ Rt 40 units**			**+ Neck**	
L	H	T	Coral +	Garnet +	Topaj	O +	Cr
24	26/28	6	40R	40b	24R	40	60
R	R	b					

In this way dozes increased corresponding to disease. At higher doze (60 or more) requirement of yellow sapphire in the form of ring starts. Ultimately Dozes culminated in the complete Sluggish patient refused by AIIMS, New Delhi.

Rt

Coral	+	Garnet	+	Topaj	+	Saphire
218		450beads		36/45		7R (yellow)

Left

L	+	H	+	Tig.	+	cats eye
129		150		32/35		9r
r						

Neck				**Rt arm**		
O	+	cr	+	Pearl	+	Emerald
218		320		10		10
b		+ 60 b		r		r

After this doze (in 3 months), he started walking first on Baisakhi and then on the bicycle even.

From above discussion it is apparent that there are 3 categories of dozes here A(a) which includes either right hand only (3,5,10 and 12 units), (b) both hands (12,15,20 units or), (c) both hands + Neck (15,20,30,40,45,50,55 and 60 units). Maximum categories of diseases are covered within these dozes. Further increase in dozes depends upon the seveority of same types of symptoms. The most graveous patient refused from AIIMS was also a patient of back pain but was complete sluggish. Under such circumstances dozes used of the same characteristics but dozes increased. (B) Left + neck - usually for low B. P. Gastritis, some heart troubles and in some liver cirrhosis cases, etc. This is in brief about categories. But selection of doze (characteristics) is the main thing. Increase of the doze depends on minute observations with tools, depending upon following factors.

(1) Reversal of healing (after giving doze patient reports relief but again after 48 hrs it reverses both in report and tools). The doctor should be more dependent on tools but patients be asked repeatedly and definitely. As under ailment patients become psychological (a bit) and important at the same time.

So, during increase of the doze doctors insight counts much with the tools. Since withdrawal of doze is main quality of this treatment as because it is not swallowed, however it becomes time taking and slight troublesome for patients. After withdrawal of any doze patients must be left untouched for 48 hrs. In very emergency situation also 2 - 4 hrs be left after withdrawal of all dozes endorsed. After 48 hrs of rest the same doze is again endorsed. Thus increase in doze depends upon the demand of the body. Demand of the body means in its disease symptoms and general body language. So, in any symptom dozes may go up. For example, in headache - it may be relieved in one doze 10/15/20 units or may go upto 45/60 units as if headache is due to migraine. This may be the case in all symptoms. So, for any patient after thorough diagnosis a prescription is written. Usually that comes true but sometimes may require more dozes. This depends upon the skill and experience of the A. G. doctor. So, the dozes under discussion go like this:

60 unit -	60 + 60/90 + 30/36	L + H + T	O + crys
	coral garnet Topaz	36 42 9	60 + 90
	Rt	Left	Neck

The ring of precious stone is suggested if diseases are inhereditary anyway.

So, saphire yellow is added in ring i.e., in the index finger (5/7 ratti).

Such characteristic dozes in this empirical way go up to 70 units → 80 units + 90 unit → 100 units → 110 units → 120 units → 130 units → 140 units → 160 units → 170 units → 180 units → 190 units → 200 units → 210 units → 215 units → 220 units. Always dozes should be increased very cautiously when symptoms remain unchanged or reverts and patient is not appearing cured. In that case patient be left to settle after a fortnight dozing. But patient must be Left at stable doze if it is not complete. As if 20 unit is not doze of cure then it is with full L + R + neck must go to settle. 30 is not stable, 40 may be stable, 45 is also stable, 60, 90, 100, 120, 140, 160, 180, 200, 215, these are stable dozes where all stones according to formula are endorsed no fraction of any doze remains. This is the criteria of stable doze.

Now, there is another category of dozes (B) pattern. If the patient falls under (B) pattern he/she can't tolerate (A) pattern dozes. They will show negative results. Eg. the patients of rheumatic heart, the upper (A) initial doze will aggravate his/her symptoms. The patient may develop fever. Whenever, doctor finds negative results he should judge his dozes and diagnosis first. This is all about pattern (B) dozes and range. Now doctor should apply his wit, otherwise next doze should not apply. Initially an error may be committed and gradually will be rectified. The pattern A is usually applicable in normal patient with normal pattern of B.P., temperature and heartbeat. Here only few precautious be maintained:

1. Weight of dozes should be exact.
2. All initial readings should be exact such as temperature, B. P., heartbeat.
3. In cosmograph Jupiter should not be so weak (from mount if no cosmograph). No marked gas trouble allergic effects. Body temperature must be taken from side. If blood circulation problem is there then from both sides temperature must be taken after breakfast and never empty stomach.

Doze pattern (B) – gas trouble, allergic effects symptoms in this type of dozing. Dozes of right side are completely eliminated. Dozes are endorsed in left and neck.

Patients characteristics – Patients of weak Jupiter (either from cosmograph or Mount), weak body, Low B. P., high palpitation. gastritis condition i.e., weak lever, high heartbeat, rheumatic heart, increased body temperature. Saturn and rahu are strong unlike the patients of pattern (A).

Patients of such characteristics are regulated in left hand and neck. Even then first two dozes be endorsed with sign of interrogation whether this is suitable or not. If temperature, B. P., heart beat all recovers patient feels good then go ahead switch it on.

This pattern goes to some limited scale i.e. fewer number of dozes are required.

Range of Dozes - pattern B:

(a) 5 U - Left Neck

L	H	T	Onex	crys
5R	5R	nil	10 +	10
			beads	beads

(b) 6 U - Left Neck

L	H	T	Onex	crys
6R	6R	1b	12	12 + 1 + 1 + 1 = optional

(c) 7 U - Left Neck

L	H	T	Onex	crys
7	7	1	14	14 (3 optional)
R	R	b		

(d) 9 U - Left Neck

L	H	T	Onex	crys
9	9	1	18	18 + 3 (op)
R	R	b		

Additional crystals are given in dozes for making balance of temperature after endorsing proper doze. Crystal if required must be added one by one to bring temperature 97°.

(e) 10 U - Left Neck

L	H	T	Onex	crys
10	10	2	20	20 + 5 (op)
R	R	b	beads	beads

(f) 12 U - Left Neck

L	H	T	O	crys
12	12	2	24	24 + 6
R	R	b	beads	beads

(g) 15 U - Left Neck

L	H	T	Onex	crys
15	15+(2)	3	30	30 + 5 (op)
		b	beads	beads

Extra dozes in hessonite and crystal will balance temperature if goes high/low. Higher dozes of hessonite are sometimes constipative.

In this pattern, dozes endorsed upto 60 units where required. This is very simple that a doctor should endorse a doze and read the body parameters through tools, if not become balanced next doze or extra beads given are endorsed. The pathological test of such patient must be suggested at least of haemoglobin. Heart patient must not be regulated by higher dozes of (B) pattern. For some fate parameters few dozes are approved as IIT/high competitions. Following tested dozes are advisable to good examinee. It balances fate upheavals, or fate misbalances.

R + L - 10U dozes, if not balanced (B. P. 120/80, 110/70, temp 97°) then should be increased to 12U/15U with yellow sapphire 5 ratti in gold - Index finger. In marriage problem also this doze is suitable but for otherwise hankering after wealth/ irrelevant achievements not encouraged. These are cheatings/ fallacious.

Endorsement of Dozes

After proper weighting and beading in sequences mentioned below for right and left arms, wearing of dozes started from the left arm and then right arm after a pause of at least 5 min. They were told to stay for 30 minutes. After 15 minutes they were again checked up and told to come after 48 hrs. If the case was a bit serious, the patient was told to come next day also (24hrs.)

Left Hand and Right Hand:

Rt			Left						
C +	Garnet +	Topaz	L +	H +	T +	Cats eye +	yellow saphire	O +	crys
80R	80 beads	34+2	48/52/56	56/60/64	12	7	6-7	80	120
		ratti	ratti		beads		ratti	b	b

This is maximum doze if haemorrhage has not occurred.

It includes cancer (carcinoma), lymphoma, osteo carcinoma and liver cirrhosis except in liver cirrhosis where B pattern of dozes also may be suitable. (A) pattern is regulated upto 80 units (maximum) but good yellow sapphire essential upto 7 ratti.

In rhumatism also pattern (B) Suits upto 30/40 units. Yellow sapphire also needed due to genetic nature

Rhumatoid arthiritis - Pattern (A) is regulated maximum upto 45 unit but yellow saphire needed 5/6 ratti.

Allergy and Asthama: Pattern depends upon body, high B. P./Low B. P., wt of body. If there is high B. P., then pattern (A) will be suitable otherwise (B).

Peculiar dozes are suitable here:

Rt	Left		
10 unit doze -	C + Garnet + Topaz	L + H + T	+ Cats eye
	10R 10 (may or- (may be- may not) increased) suits	3 3.5 1b	(3 to 7 ratti)

Yellow sapphire 5R (if chronic) be added in ring. Same constitution dozes be increased upto 27U/or more. In some weak patients with low B. P. and gastritis Pattern B (Left Neck) dozes be endorsed.

Complete doze pattern of asthama will be described later on. However 2 dozes described for low intensity patients here.

1st Doze - (Given above)

IInd Doze - add as described (same)

Doze May further increase add as described (same)

Further Doze - as described (upto 27U) (same)

Rt

Sometimes - Same

Precaution - Same

IInd Doze :-

Rt			Left		
C +	Garnet +	Topaz	L +	H +	T
15r	15+5	may or may not 5+2.5	5R	5/6R	2 beads

SUMMARY OF DOZES DESCRIBED UNDER PATTERN A

No. of	Patient	Unit		Rt			Left			Nexk
Dozes			C	G	Topaz	L	H	T	O	Crys
1st doze	Infant	3U	3r	3b	nil	nil				
2nd doze	Babies	5U	5r	5b	nil	nil				
3rd doze	Adult	10U	10r	10b	nil	nil				
4th doze	Adult	10U	10r	10b	nil	3r	3.5r	1b		
			(with breathlessness)			3r	3.5r	1b		
5th doze	Adult	10U	10r	10r	nil	3r	5r	1b		
6th doze	Adult	10U	10r	10b	nil	5/6 R	5/6 R	1b		
7th doze	Adult	10U	10r	10b	5r					
8th doze	Adult	10U (full)	10r	10b	6r	5 R	5.5	1b		
9th doze	Adult	12U	12r	12b	7.2r	7.2	8	1b		
10th doze	Adult	15U	15R	15b	10R R	9/10 R (Eosonophilic)	11/12	2b	nil	
11th doze	Adult	20U	20R	20b	12R	12R	12/ 13/14 r	3 b	20 b	20 b
12th doze	Adult	20U	20R	20R	12R	12 r	14 r	3 b	20 b	20 b
13th doze	Adult (sugar Patient)	20U +2/3	20R beads	20b pearl	12T	12/13	14/15	3	20 (with/ withought)	30
	(Normal)	20U	20R	20b	12T	12/13	14/15	3	20b	30b
14th doze	Adult	30U 25U doze/may be tried with same ratio.	30 R	30 B	18	18 R	19.5/ 21 R	4 b	30	45
15th doze	Adult	40U semi st.	40 R	40/45 B	24 R	24/26	26/28	6	40	60
16th doze	Adult	45U Stable doze Garnet 1 by 1 to maintain temp. if required (14/15/16 and 20)	45 R	45 B	30 R	30	32	6 b	45	70

No. of	Patient	Unit		Rt			Left		Nexk	
Dozes			**C**	**G**	**Topaz**	**L**	**H**	**T**	**O**	**Crys**
17th doze	Adult Not stable	50U R	50 b	50	32	30	35	7 b	50	75
	N.B.- fine adjustment in the dozes of L&H of .25 to .5 R may take place. whether, the dozes may be used for slightly serious patients.									
18th doze	Adult/old	55U	55 R	55 +R	32/33 R	32/33 R	37/38 R	7 b	55	82/83
19th doze	Adult/old	60U	60 +5+5 R	60 B	33/36 R	36/39 R	42/45 R	9 B	60 B	90 B
20th doze	Adult/old	65U	65	65 + 10 or more	34/35 R yellow saphire may be added (5.5R)	42/46 R	45.5/47 b	9 b	65 b	97
21st doze	Adult/old	70U	70 + 10 or more	70	35/36	42/45	49/51 b	10 b	70 b	105
22nd doze	Adult/old	75U	75	75 + 10 or more	35/36	45/47	52/54 b	10 b	75 b	112/113
23rd doze	Adult/old ailing	80U	80	80 + 10 - 20	36	48/50	56/58	12 b	80 b	120 b
24th doze	Adult/ailing	85U	85	85 + 10 - 20	36	51/54 R	56/59 R	12 b	85 b	128 b
25th doze	Adult/ailing	90U	90	90 +	36 +1 - 2	54/56 R	63/65 R	13 b	90 b	135 b
26th doze	Adult/ailing	95U	95	95 + 47	37	57/61	67/69	13 b	95 b	142 b
27th doze	Adult/ailing	100U	100 + (may be increased 1 by 1)	100 + 50	37	60/65 R	70/75 R	15 b	100 b	150 b
28th doze	Adult/Ailing	105U	105 R	105 + 55 b	36 +2 (max.)	63/67 R	75/77 R	15 b	105 b	157 b
29th doze	Adult/Ailing	110U	110 R	110 + 55	36 +2 R	66/71	77/81 R	16 b	110 b	165 b

No. of	Patient	Unit		Rt			Left		Nexk	
Dozes			C	G	Topaz	L	H	T	O	Crys
30th doze	Adult/Ailing	115U	115	115 + 57	37	69/73	81/86	16 b	115	172
31st doze	Adult/Ailing	120U	120	120 + 60	37	72/78	84/90	18 b	120	180

Asthma Patient II doze:

Doze may further increase:

Rt			Left		
C +	Garnet +	Topaz	L +	H +	T
20	20+5	may or may not 10r	6.6r	6.6/7.6/8	3

Further Doze:

C +	Garnet +	Topaz	L +	H +	T
27r	27+6	may or may not 13/13.5	9	9+3	4

Rt

Sometimes normal dozing takes place upto 30-45 units.

Precaution - Such allergic patient should not touch stones other than dozes.

Cancer is due to the interactions of :

Sat - MA + organ

+R (for malignancy) (ME/V/others)

Malafic SAT - MA - forms Pro-oncho and when R comes under interaction then organ becomes malignant. It is Lamented that due to lack of facilities/patients of hydrophobia, titanus, AIDS could not be regulated but new practioner may try with dozes and patterns.

Precious stones like yellow sapphire, diamond, emerald may be required as the case may be. Genetic diseases are also regulated with (A) upto 60-80 units but yellow sapphire is must (5-7 ratties).

Our slogan now appeared quite relevant in the eyes of modern science that cosmic rays are causes of all genetic phenomenon, already discussed in mechanism portion. Here we will discuss that cosmic rays are the causes of all human diseases. Really modern medical science while ascertaining the causes of diseases appears ambiguous which leaves certain question mark. Medical science emancipated food habit and life style as main causes. But when small children or infant suffers from cancer, diabetes, medical science keeps mum. There is apparently no reply. When asked from doctors, they frankly tell that we are responsible to treat only may be on symptomatic basis, so we are not supposed to reply how and why? They on other front of diseases ascertain bacteria and virus as causes. My first discussion with the doctor of AIIMS was based on the same fact. He claimed that we apparently see that pneumococcus bacteria is the cause of pneumonia how can you say cosmic rays? Then it was categorically replied that if without adding antibiotics patient is cured successfully by altering the effect of bad rays. Actually, bad rays when grip the patient they bring some biochemical changes in the body of the patients, preparing suitable media/protein for the growth and virulence of that bacteria if rays are altered, medium of growth reverses and hence bacterial population vanishes thus the patient becomes cured.

It will be possible with a kit of semiprecious stones, few fishing strings of zero number, thermometer, B. P. instrument, fine electric wire and weight meter to start self and family treatment. Just to avoid family Doctor for normal and seasonal ailments even small chronic ailments. But it is noticed that several families form queue on doctors clinic and spend heavy recurring expenditures. Whereas the cost of stones once purchased appears negligible as they are not consumed rather used in patient to patient mission. However, this is for information that self-dependent routine treatment (seasonal, a bit chronic, long duration treatments are easily possible to treat and avoid family doctor), terminal and big chronic diseases may also be regulated but for that training is must (dozes and methods for their regulations will all be illustrated).

Doze to increase immunity of the body up to range i.e. cough, sneezing, fever at frequent intervals, and bowel trouble etc. remain subsided with 10, 12 or 20 unit dozes, through which one becomes balanced in the maximum instabilities. Loss of concentration are also regulated by primary dozes. Arrogance, bad habits, laziness may be regulated by dozes up to 40-45 units depending upon the duration and frequency of the trouble. If there is genetic link then yellow sapphire may be added.

As elaborated previously 12 pairs of cranial nerves supply and control the organs same as 12 *rashies* and 27 important spinal segments are attached to and control the same organs as depicted in constellation. So, similar pattern in human physiology and bramhand is observed. Following is the account and extent of similarities shown by constellation and spinal segments.

T 9 Kidney — Purva phal. **Ven**

T 10 Small intestine — Jyestha. **Mer**

T 11 Appendix. caecum, colon. Ase — Swati **Rah**

T 12 Large intestine — Satvikha **Rah**

L 1 Large intestine — Anuradha **Sat**

L 2 Testes, ovary — Purvakha **Ve**

L 3 Bone marrow, femur — **Mo**

L 4 Bone marrow — Purvabhardra. **Jup**

L 5 Knee, tibia — Uttarbhardra. **Sa**

S 1 Foot — Rebti. **Me**

S 2 Rectum anal canal — Chitra. **Ma**

S 3 Uterus, prostate, genitella — Ghanistha. **Ma**

S 4 Fellopian tube urinary BL. Vasdifference — **Mool ke**

C 1 Head, nose, sinus, — Mrigsira. **Ma**

C 2 Brain, skull, ear — Mrigsira. **Ma**

C 3 Head, brain, temporal region — Ashwani

C 4 Mouth, lower jaw — Bharni. **Ven**

C 5 Neck — Pukhya. **Sat**

C 6 Arm — Ardra. **Rah**

C 7, C 8 Arm, hand wrist — Aslekha. **Mer**

T 1 Thymus gland — Punerv. **Jup**

T 2 Chest rib — Magha. **Mo**

T 3 Lung — Rohini. **Mo**

T 4 Heart — Uttara Khar. **Su**

T 5 Upper part abdomen, stomach — Kritika. **Su**

T 6 Lever spleen — S Uttarphalguni. **Su**

T 7 Do — Hast. **Mo**

T 8 Gall bladder, pancreas — Visakha. **Jup**

Following is the account and extent of similarities with cranial nerve and *rashies*.

Organogenesis — BBB under analogies of cranial nerve — Rashi Cr.

N	**Organs**	**Rashi Own**
1st P	Nose (internal layer)	Mesh. **Ma**
2nd P	Retina	Brikh. Ve
3rd P	Muscle of eye ball	Mithun. **Me**
4th P	Muscle of eye ball	Kark. **Mo**
5th P	Pinna, lower face, teeth, Gum, eyebrows forehead, Nose (external) rest part of eye ball	Singh. **Sun**
6th P	Muscle of eye ball	Kanya. **Me**
7th P	Internal ear, muscle of face, neck, tongue Salaivary gland	Tula. **Ve**
8th P	Internal Ear	Brischik. **Ma**
9th P	Tongue, pharynx, pallet	Dhanu. **Jup**
10th P	Larynx, trachea, Lungs, Oesophagus, Stomach Intestine, heart, gallbladder, Kidney, Lever, Pancreas Spleen	Makar. **Sat**
11th P	Neck, Shoulder	Kumbh. **Sat**
12th P	Tongue, reflex action spinal	Meen. **Jup**

For regulation of diseases, organogenesis is must, because it refers sometimes to organs or biochemical products, thus their translation to different interactions will make the process easier. So, we have discussed about one pattern which resembles much with the Body physiology. Organs attached to the cranial nerves are more or less similar to those referred by Rashies. Again those controlled by spinal segments are approximately same to those denoted by Constellations. More over both patterns marvellously analogous. There are 12 rashis and 12 pairs of Cranial Nerves. Next there are 27 constellations 27 important spinal Segments. Hence organogenesis is completed and translated.

1. According to number of houses of cosmograph rt from 1st (i.e. Lagna)
2. According to cranial nerves..... Rashi
3. According to spinal segments, constellations.

1\. Lagna (1st House) — +Aries — = fore Head

2nd House — +Taurus — = mouth, throat, chicks

3rd House + Gemini = neck, hand, shoulder, arm, heart

4th House + Cancer = stomach, chest, kidney

5th House + Leo = main arota

6th House + Virgo = stomach, head

7th House + Libra = genitellia

8th House + Scorpio = height and reproduction

9th House + Sagittarius = high leg joints

10th House + Capricorn = ankle, knee

11th House + Aquarius = internal part of stomach

12th House + Pisces = leg

Moreover like human genome characters/genes situation is studied, similarly in this old science characters have specific sites as loci in genetics (justify genetic phenomenon displayed).

1. Saturn:

- First layer: Bone, lower teeth, large intestine, muscle

- Second Layer: agriculturist, dakpeon mi ser, head, malee, potmaker, police or police officer, fasting, saint, hilly areas, hill, non-cultivable, cave, old, graveyard field
- Third Layer: Material knowledge, views, independence, grave, self-control, laborious, patience, character, decisiveness, intellect, alertness, considerate

2. Rahu: Insomania, bowel, brain, madness

3. Ketu: Skin disease, mental trouble, appetite, disorder Important any planet is not controlling any organ fully. Any organ has many parts, so they partly control. Any organ has its main body and muscles, blood functional parts, etc. thus under control of many planets. As discussed earlier rays affect body in 3 different ways, 1, 2 are physiological layer, as apparent from discussion of the layers.

4. Mercury:

- First layer: Brain, nervous systems, speech, hand, artistic approach
- Second Layer: Understanding memory, logic, agree-disagree, fine art
- Third Layer: School, college, science, scientific endeavours, literary publications, post master, businessman, green effect of mercury, mithun. heart

5. Sun:

- First layer: Main arota, blood circulation, eyes, small capillaries, tooth, ear
- Second Layer: Greatness, self-control, views, balance, utmost, capability
- Third Layer: King, minister, Leader, military, invention, archaeologist

6. Venus:

- First Layer: Neck, kidney, face colour, hair, beauty, body, control, organ, sex
- Second layer: Affection, rest, beauty concept love, cleanliness, working capacity, idea

- Third Layer: Beautiful, ornaments Dance, music, amusing artistic, decoration, arrangement, motherhood

7. Jupiter: Three tiered characteristics of planets and Stars of that planet traits. 1. ... Jupiter

- First layer (saririk)..leg, Thigh, Lever, Digestion, Bloodpressure, Nerve.
- Second Layer (Aatmic).Good nature, benevolence, beauty lover, strength, devotion, organization, wisdom, intellect.
- Third layer (Aatmic), temple, worshipper, minister, judge, teaching institute, university donation, affection.

8. Mars:

- First Layer-.external head, nose, chick, contagious disease, wound, operation, pain, blood defects
- Second layer-bravery entreprenureship, consistency, self reliance. anger, fighter, attitude, impression
- Third layer- military, doctor, professor, engineer, chemist, carpenter, blacksmith, machine works, manson, play and play material dealer/representative.

9. Moon:

- First layer-stomach, digesting, power, breast, uterus, vagina, eyes female genitalia
- Second Layer- sensitivity, internal instinct imagination, alertness, homely, Greedy, nervousness
- Third Layer-white colour, ship, fish, water liquid, nurse, maid servant, food, silver, blue Cosmic rays are causes of human diseases - In this concern, one example is discussed with AIIMS and described already, e.g. pneumonia.

Another example is that of E. Coli, B. Coli. They remain in the body, helping in faeces discharge, but some patients suffer, Urine infection, fever etc. It is remarkable that contact of rectum and penis always takes place in every individual, but very few of them report virulence symptoms of disease. There is no reply by medical science. Same reason prevails here also.

The patients when gripped by bad interaction of rays, they produce some biochemical products/protein which is serving as growth medium for the growth of bacteria in urinary tract, bacteria flourishing then well there. When rays are altered, virulence seizes thus Patient recovers. In the cases of liver cirrhosis medical science is of the opinion that it is due to wine, cigarrete, tobacco, but when patients appear without taking anything of them. Then doctors keep mum. So, after meditating upon, it is apparent that food, bacteria, viruses are simply stimulatory. Main causes are cosmic interactions (Primary). There are many such cases coming across in our life but we ignore as infants suffer from terminal diseases (blood cancer, blood sugar). So, above assertion is beyond doubt correct. So long cosmic rays was not under scientific preview, there was a confusion.

In left and rt dozes some stones were weighed but garnet, tiger and onex crystals were in count decided in accordance with disease and power of tolerance, weight etc. of the patient. Diseases wise constitution of dozes is given in detail. However one thing is essential to be mentioned that weak patients with low B. P. and weak Jupiter are advised altogether different doze. That is in left arm + neck as mentioned previously. Detained patients were first called and after minor queries prescriptions were served. In chronic and complicated cases precious stone might have been suggested along with semi-precious dozes. The patients were asked to stay in nearby places because some of them coming from outside the town sometimes from very remote places. Normal patients were asked to stay 7-10 days. Chronic patients were also suggested to stay not more than a fortnight. Everyone was suggested to come again. Chronic patients may be called maximum up to 3 months. No treatment was beyond that. Usually time in normal patients were calculated in respect of dozes which were endorsed every 48hrs. After completion of dozes, patients were detained for observation upto next 48hrs and then told to visit once at least after a fortnight. As dozes are endorsed, ailments vanish. Dozes were endorsed exogenously nothing to eat. Even it is named as dozes, because it is immediately effecting interiorly more quickly than dozes to eat.

After due process, the voluntary society was formed. Members started joining hands. Their numbers gradually increased. Few tools were purchased all from pocket like thermometer, wt. meter, the most durable threads usually used for fishing net. Small balance used for stone weighing, its weight box, register for recording etiology of the patients, B. P. instrument, magnifying glass, big carpet, pillows, benches, etc. Patients were called one by one. First of all, one assistant entered the details of cosmography, if available planetary position and period. Then other recorded etiology, weight, pulse, temperature, special if any. Then patient brought before the director. He used to observe planetary position and interactions in mounts of the palm of the patient. Then it was compared with cosmograph readings. Now B. P. was recorded and heartbeat, sound were recorded, patient was asked his ailments, i.e. any pain anywhere bowel, wrist, etc. In case of an old patient, previous treatments records were observed. In chronic cases few pathological tests were suggested especially hemoglobin. Then instantly prescription of combination of stones (semi-precious) was given. If the patient appeared complicated and chronic, case was (detained for a day to study the case patiently.)

Astrogenetic mechanism developed a powerful therapy of human diseases. For the exercise of the same therapy ICMR New Delhi suggested to form an Ethical Committee. Comprising of four members: 1. Retired Judge 2. High Court Pleader. 3. Physician. 4. *mukhia* of village panchayat. Thereafter a voluntary committee named Astrogenetic Society was formed with a view to serve humanity (ailing mankind) and in this way a journey of Manav Kalyan started, atleast continued for 25 years. Vigorously day and night.with a slogan and mission in mind. "Service to man is service to God", i.e. with *manavtavadi dharm*. In this mission, like-minded people started to join voluntarily, and started giving time. In this way 5-6 full time members joined hands.

The mission is very important and special because while rendering services to mankind a vital theme discovered, i.e., the "control system of nature and scientific development of events in the cosmos. Moreover, a clear cut idea of science and *adhyatma* emerged. The confusion regarding God may also be satisfied. More elaborately, Scientific analysis now may be made why she is my wife, Sister why he is my father, brother, neighbour, etc. from this platform one can reply every bit of questions. We see that when Doctors asked to explain causes of diseases he shirks and tells to take medicines only. But under the most minute layers of knowledge, there is answer of every query. Even this Science cannot give radical cure to every rejected and dejected patients as they already have lost many things which can't be re-generated however, it is giving them treatment, more than any pathy in simplest way at lowest/no costs. The materials used, remain as such. This pathy may be used - for self treatment, family treatment and it may also be taken for lively hood. But humanity must not be jeopardised anyway because here we handle most sacred Godlike force (cosmic rays). So, honesty must be there.

Range of Diseases

For pattern (A) dozes:- Any trouble of infants (upto 5 yrs) 3 unit doze, any trouble of medical range upto 10 years children 5 unit single arm doze. This single arm doze is suitable for the patients of left cold characteristics. Otherwise should also be balanced with left (Lepis 2/2.5r - Hess. 3/3.5r). This doze is suitable for patients suffering from cough, fever etc. Here balances may be done by increasing garnet one by one.

- 10 unit doze, with or without left: If left added, Topaz should be added in rt. The same for 12 unit. These dozes are very useful for family doctor range diseases - cough, Eosinophyll, fever, headache defect, eye pain, nose, ear defects, throat sore, etc. all in low intensity or in children up to 12 years. It includes pain in stomach, bowel disorder for children with normal lever.
- 15 unit onwards i.e., 20 unit are recommended for adults also in cough, headache, pains bowel disorder, concentration, immunity.

- 20 unit full doze, i.e., 20 coral + 20 b Garnet + 12 r topaz (R) 12 Lepis + 12/13/14 (Hessonite) + 3 tiger (L) and 20 Onex + 30 crystal (N) is a very stable doze for medium catagory also.
- 30 unit doze is rarely stable in pains (back, abdomen, head and) digestion otherwise 45 unit is stable for family doctor range diseases patients with seasonal ailment within which slightly chronic trouble persisting. Now onward dozes increased when diseases remained as such but for terminal diseases also dozes may be increased if the body demands. Increase in the doze should be basically on the demand of the body. That demand should be transcripted from tools and cosmography. If patient possesses cosmograph the plight of diseases be concentrated. Always in increased conditions of dozes atleast one precious be added preferably Saphire (yellow). All types of medium ailments should be subsided upto this doze. For paralysis, Cancer and other dreaded diseases, this doze is threshold. Further increase may be required by body weight and ailments. In that case, maximum range has been displayed i.e. upto 60 - 80 units which can be tackled easily. Main thing is selection of the1st doze either in (A) or (B) pattern and supplementation of additional stones.
- 20 to 45 unit dozes are sufficient for B.P., Sugar, Anaemia and pains. In case of Sugar (diabetes) 20 unit balanced doze requirement takes place with Pearl in rt (3 to 5 ratti). That combination may increase upto 45 units as the case may be in the above dozes, usually precious stones were applied to the patients regulated with R + L... 60, and above.

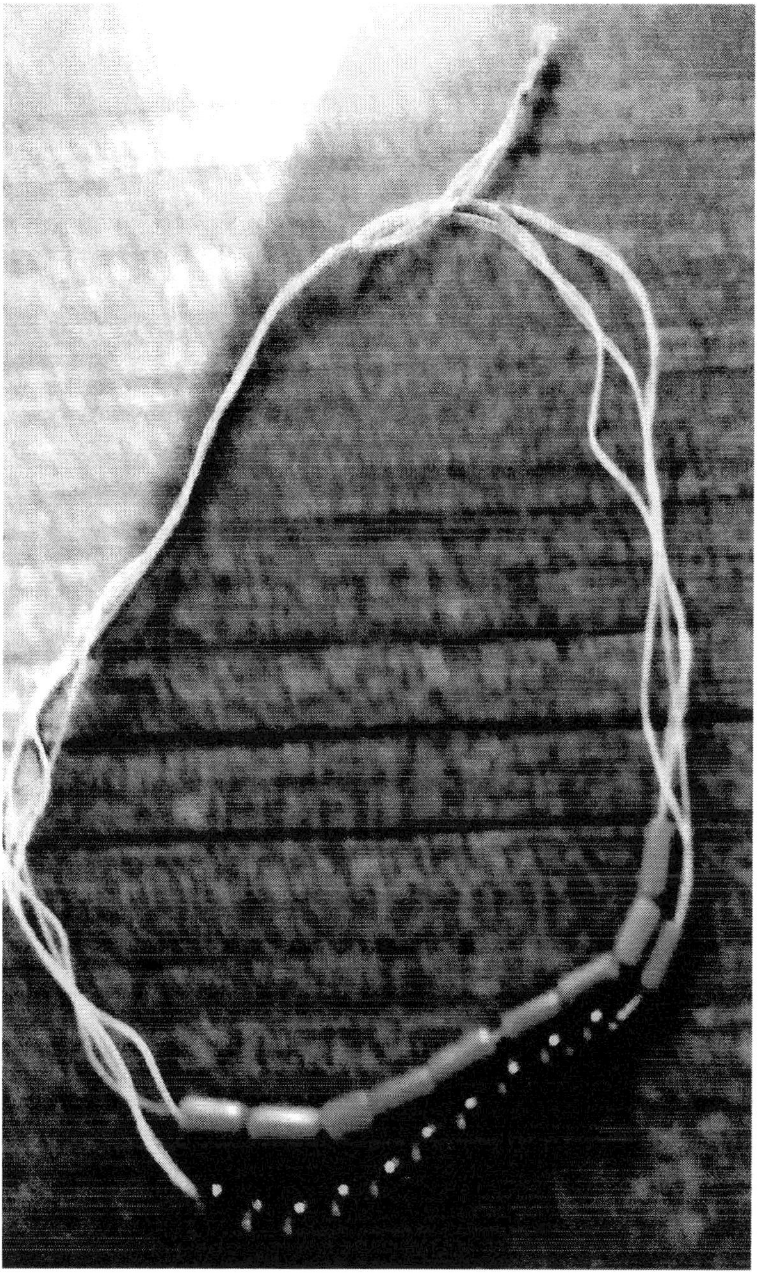

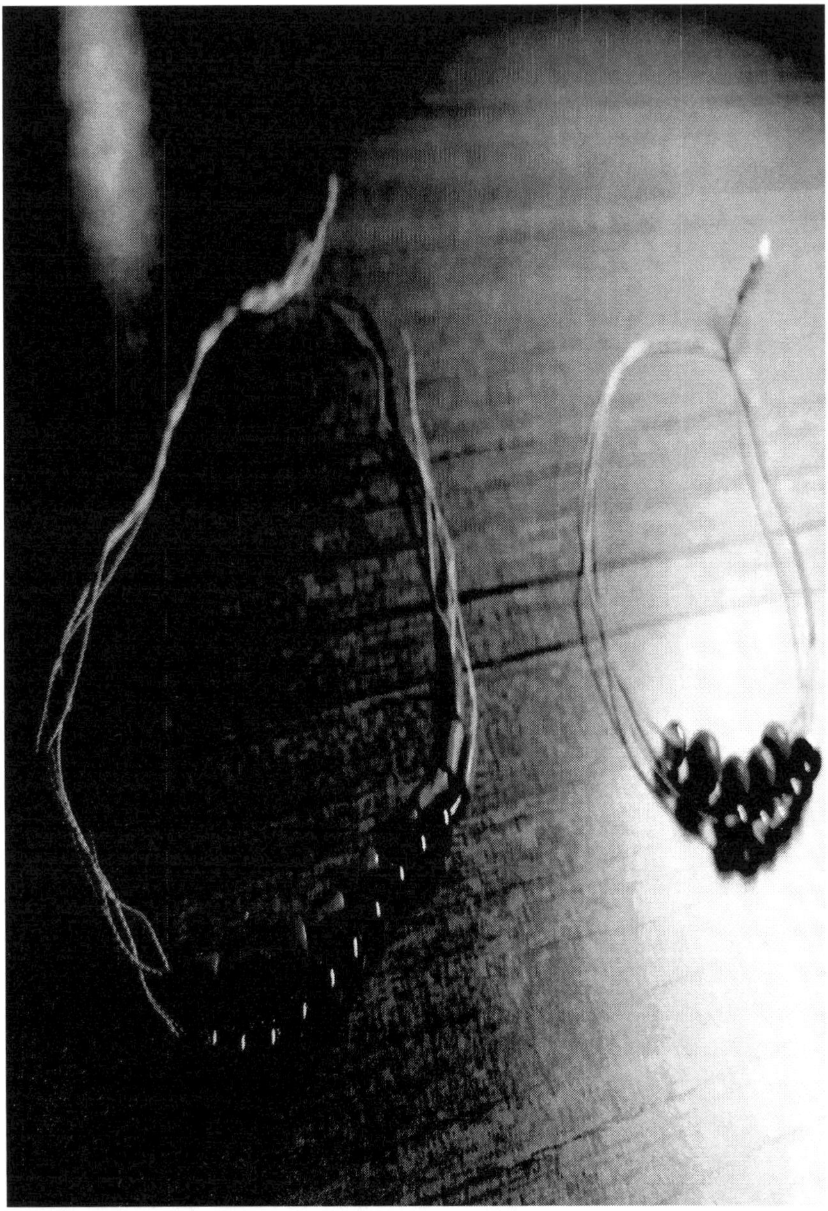

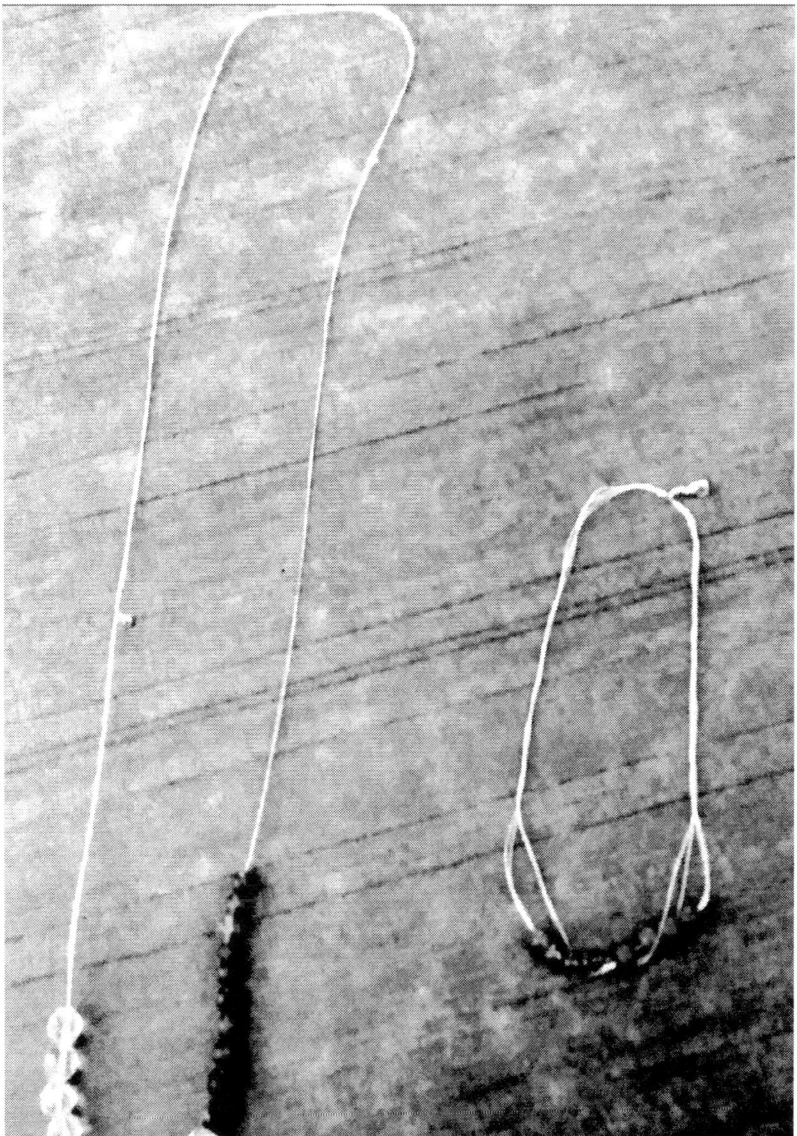

Chart Group (A)
(Orthopaedic & Neurology)

S. No.	Patient's Name	Diseases	Previous Treatments	Medical Reports after Astrogenetic regulation	P Value
1.	Sri Umanand Mishra	Acute Arthritis	Dr. Lt. K. Jha- Patna Dr. B. Mukhopadhyay. AIIMS, New Delhi	Enclosed do	
5.	Dr. Arun Kumar Das	Acute Spondo-litis	Calcutta & Darbhanga	do	
10.	W/o M.K. Sinha	Arthritis	Nil	do	
14.	Mr. Purnendu Mishra	Arthritis	Nil	do	
19.	W/o Sri R.N. Jha	Sciatica Acute & Asthama	Dr. U.K. Jha, Patna (Neurologist)	do	
20.	Prof . Chandra M ohan Jha	Paral ysis, loss control over eye-lid	Local and AIIMS	do	$p>0.01$
21.	Sri Indradeo Jha	Paralysis & Parkin-son	Local and AIIMS	do	
24.	Prof. Nripendra Nath Jha	Spondo-litis	Dr. B. Mukhopadhyay Patna	do	

Chart Group (B)
(Genetic)

S. No.	Patient's Name	Diseases	Previous Treatments	Medical Reports after Astrogenetic regulation	P Value
2.	Sri Birendra Kr. Shrivastava	Liver . Cirhosis	AIIMS, New Delhi IGIMS, Patna, Kothari Medical Calcutta	Enclosed	

S. No.	Patient's Name	Diseases	Previous Treatments	Medical Reports after Astrogenetic regulation	P Value
3.	Sri Chandra Madhava Jha	Sreoma	Dr. Abdulbari Khan Darbhanga, Tata Memo rial, Bombay	do	
4.	Suman kr. Jha	Mental Retardation	—	do	
15.	Prof. Amarnath Roy	Heart, Valve effected (Genetic)	Dr. A.K. Thakur, Patna Safdarjang Hospital, New Delhi, IGIC, Patna AIIMS, New Delhi.	do	
16.	Smt. Dulari Devi	Asthama	Dr. T.K. Jha, Muzafferpur	do	
17.	Pankaj Sarraf	Leucoderma & Eiplepsy	—	do	$p>0.001$
22.	Vikas Kumar	Mental	Calcutta & Ranchi (Dr. K.K. Sinha)	do	
25.	Abhishek Kumar	Muscular Distrophy	AIIMS, New Delhi	do	
32.	Sumit Kumar	Muscular Distrophy	Dr. U.S. Jha, Darbhanga AIIMS, New Delhi	do	

Statistics Chart Group (B) (Genetic)

S. No.	Patient's Name	Diseases	Previous Treatments	Medical Reports after Astrogenetic regulation	P Value
6.	Miss Nandini Kumari	Myopia	Physicians of Patna	do	
7.	W/o Late Lutan Jha	Parkinson	AIIMS, New Delhi	do	
8.	Sri M.K. Sinha	Myopia & eye bleeding enternally	Dr. Anjani Kishore pd. Patna	do	

9.	Smt. Meena Devi	Migraine	Dr. M.Thakur Dr. Mrs. Arti S. Niranjan (Patna)	do	$p > 0.001$
12.	W/o Girija Nandan Singh	Heart & others	–	do	
13.	Prof. Prabhat Kumar	Physio-logical	Darbhanga, Patna, AIIMS New Delhi	do	
18.	Miss Putul	Heart	AIIMS, New Delhi	do	
23.	W/o Laldeo Jha	Physio-logical	Nil	do	
26.	Sri Uday Thakur	Lever enlarge-ment & others	Local Doctor	do	
27.	Sri Harishankar Thakur	Ulcer, Dysentry etc.	Local Doctor	do	$p > 0.001$
28.	Sri Indrakala Devi	Gastritis & others	Local Doctor	do	
29.	Smt. Urmila Jha	Prolapse palpita-tion etc.	Local Doctor	do	
30.	Smt. Preety Thakur	Eosin-ophyll Headache etc.	Local Doctor	do	
31.	Smt. Nagina Devi	Leucor-rhia pain, recurrent fever etc. heart	Local Doctor	do	
11.	Smt. Puspa Devi	Heart	Dr. A.K. Thakur	do	

List of Publications and Others

Scientific Paper was only published by Dr. C. N. Jha but all others automatically published either through Patients or through reporters, of print media.

1. NLO - Bulletin
2. Scientific Paper published in Journal Evn. and Ecology 10(1) : 117 - 123, 1992 Effects of some Naturally occuring chemical compounds on Human diseases (Allergy and others).
3. "एस्ट्रोजेनेटिक्स" कादम्बिनी — 1994
4. हिन्दुस्तान — 15 अप्रैल
5. Times of India "Miracle cure" 15 Jan 1995
6. Time of India "Stone researcher claim................ 15 Jan 1995 (o.P.ed).
7. Indian Nation, Patna - 14 July 1995 - Astrogenetic cure....
8. "गांडीव" ऊ0 प्र0 दैनिक "लीवर सूत्रण रोगी.......... 21-1-1995
9. आर्यावर्त पटना, 13-जून 1995 - रोगों की चिकित्सा में......
10. आज, 22 अगस्त 1998 - विविध रोगों.....
11. दैनिक जागरण - 18 अक्टू 2000 - रत्नों से लाइलाज.....
12. "आज" 19 अक्टू 2000 - रत्न पत्थरों का मानव चिकित्सा....
13. हिन्दुस्तान - 3 जून 1995 - ऐस्ट्रोजेनेटिक सोसायटी
14. आज - 19 दिसम्बर 1995
15. AIR - 20/8/1991
16. AIR - 21/11/1991
17. AIR - 18/11/1995
18. Presceiption Format used by Society for Patients.
19. Few Samples of formats used (B) Declaration formats (few samples) (C) undertaking formats used by patients (few samples) (D) Photograph of Patients diary still preserved.

NLO

bulletin

CONTENTS

Power of Gems

...ing to Professor C.N.Jha ofC.M.Sc.College, Darbhanga ... is the oldest of all branch of sciences for ... and construction of humanity at large . According him there is enough scope in all walks of ... keeping in view the curative properties of gems ... days immemorial it has been used at the ... helplessness and has given splendid results. ... the power of gems have snatched many from ... clutches of death. According to him physici... every creek and corner of this Universe can be ...tted from the knowledge of astrology because ... have cured so many where medicines have failed ... paralysis, gastric ulcers, chronic insomania, can... spondilitis, heart attacks, epilepsy and likewise.

... he is of opnion that each gem has a perennial ... of one specific rays which never disappear ins... of consistant use for several year. He has given ...stification for his above diagnosis that in old days ... by kings through out the world had apetite to ...serve gems for purpose they desire.

...days Prof C.N. Jha is seriously engaged to find out ... real source of power of gems in curing diseases ... that the use of gems may be boon and not bane for ... humanity at large.

Environment & Ecology 10 (1) : 117—123, 1992
 ISSN 0970-0420

Effects of some Naturally Occurring Chemical Compounds on Human Diseases (Allergy and Others)

C. N. JHA

Department of Botany, C. M. Science College, L. N. Mithila University, Darbhanga 846004, India

Abstract

Effect of some naturally occurring chemical compounds such as coral, rock crystal, pearl garnet, topaz, lapis, lazuli, hessonite, American diamond and tigers eye has been studied on allergy and other diseases of human patients. The doses and combinations of these compounds for the proper effect were administered phyically on body surface of the patients. Its regulation was maintained on the basis of symptoms and the pathological investigations. These compounds may be developing by the interactions/influx of cosmic rays in the nature ; so they alter the effects of adverse rays for physiological and genetic functions. The discovery of these rays is important for life science, since we are all surrounded by these high energy particles. These rays are of great importance in biology, as they may cause physiological disorders leading upto mutations in genes.

Allergy in its literal sense means unusual sensitiveness to the action of foods, pollens, insect bites, antipathy or some undetected factors. It causes breathlessness, itching, patches, constant headache, intermittent fever to human being. A foreign agent after entering the body finds its way to the blood stream where it stimulate the production of a protective substance known as antibody. There are thousands of different types of anti-bodies, each one combats a specific type of foreign materials. The anti-body, neutralizes the invader and defeat the attack. Surplus anti-bodies remain in the blood stream and attached themselves to cell-walls. If the same type of foreign substance enters the blood stream again, it unites with the residual anti-bodies and ensure the neighboring cells. The damaged cells and the effect produced by them appear like symptoms of an allergy.

Medical Science has devoted much over these enormous troubles of human being but as yet, it miserably failed in perfect cure process. It suggests "allergen test" and regular administration of medicine for the whole-life. Sometimes specific dieting is emphasized regularly. In this way, such diseases became a source of constant distress for over 60% of human population for the whole life. This is the main reason of attention to this field for alternative perfect measures. The objective of this study is to assess the effects of some naturally occurring chemical compounds on human diseases (allergy and others).

Methods

Beads of naturally occurring compounds were prepared on the basis of effects, perfect combinations of "naturally occurring chemical compounds" were worked out. Thus $NOCC_p$, $NOCC_a$,

Table 1. Details of NOCC selected for present investigation.

NOCC No.	Name	Source	Physical characters
1	Coral	Italian Sea	Blood red
2	Rock crystal	Hyderabad, Koderma Hills, Tirchi	Colorless
3	Onex	Orissa, Jaipur, Russian Hills	Olive green
4	Pearl	Hyderabad, Japan	Milky white
5	garnet	Burma	Blackish red
6	topaz	Udaypur	Brandy color (yellow)
7	Lapis lazuli	Hajaribag Hill, Udaypur	Violet with metal
8	Hessonite	Hajaribag Hill, Ceylon, Gaya	Deep brown with blackish
9	American diamond	America	Colorless
10	Tiger's eye	Sri Lanka	Yellow with black linings

$NOCC_2$, and $NOCC_5$ were combined together and tied on right arm of the patient through a string. $NOCC_4$ and $NOCC_6$ were tied on the neck after 24 hours. $NOCC_7$, $NOCC_8$ and $NOCC_{10}$ were tied on the left arm again after same interval. $NOCC_3$ was separately tied on left arm at specified interval. The doses were increased at an interval of ten days keeping in view the internal resistance of the patient. Verbal and symptomatic reports were obtained regularly. Pre- and post-treatment blood reports were also taken. Such chemical compounds (NOCCs) have been depicted in Table 1.

Results and Discussion

Table 2 shows details of the patients of both sexes belonging to different social and cultural status. It further reveals naturally occurring chemical compounds' (NOCC) dose regulation (vide-infra) following introspection report till final recovery.

Patient No. 1/29-06 reported no sign of recovery after first dose of NOCC, however, second and third doses displayed high degree of recovery. Second dose disappeared sweating of foot (within 60 hours), back pain and cough. Chest pain shifted from right to left. It slightly increased within 24 hours and again turned to more soothing ones. Fever and blood pressure were found normal. Feeling of hot sensation gradully turning to normal was explained within 48 hours. Third dose, however, brought relief in breathing trouble within hours, the apetite became normal and chest pain vanished within a week after administering this dose. Facial expression of the patient improved a lot within 10 days after this final dose followed by increase in weight (1.5 kg) within a fortnight. The introspection report followed recovery in blood parameters through total count and differential count of patient (Tables 3 and 4). This patient showed improvements in neutrophil, eosinophil and monocyte counts in the differential test while increase in the numbers of RBC and the decrease in leucocyte counts was reported Table 4. Amount of hemoglobin increased after regulation while ESR decreased (Table 4). Differences so observed in neutrophils and eosinophils were found significant. However, it was insignificant for monocytes.

Patient No. 2/39-07 reported relief in urine congestion stool constipation and eye pain

JHA

Table 2. Etiology and doses of NOCC required for recovery of the patients.

Serial No./ reference No.	Patient	Etiology	Symptoms	Duration of Disease	NOCC No	Doses of NOCC (g) I	II	III	IV	V
1/24-06	Name Md. Ilyas Address Vigo ; Urdu Darbhanga Occupation Teacher	Age 55 years Sex Male Weight 40 kg Height 1.42 m Complexion Medium Special Deaf	Chest pain, cough, foot pain, sweating of foot, breathing trouble, blood pressure (slightly high), loss of appetite, back pain, intermittent fever.	Since 4 years	6	1.815	0.605	0.00	0.00	0.00
					4	0.605	0.605	0.605		
					1	2.904	3.63	0.605		
					5	2.999	2.42			
					3	12.1	6.05			
					2	21.78	9.075			
					7	1.694	0.968	0.484		
					8	0.726	0.726			
					10	0.968				
					9	1.694				
2/39-07	Name Kashi-nath Jha Address Kanakpur Lohna Road Darbhanga Occupation Agriculture	Age 45 years Sex Male Weight 52 kg Height 1.57 m Complexion White Special Wearing spectacles	Breathlessness, joint-pain, abnormal pulpitation, urine congestion, pain in right eye, constipation, disturbances in sleep, ill tempered, loss of concentration and confidence, Siemen discharge with urine, allergy from curd and plaintain.	Since 12 years	6	1.452	0.605			
					4	0.484	0.484	0.605		
					1	1.452	1.21	1.21	1.21	
					5	1.21	1.21	1.21	1.21	
					3	6.05	4.84	[illegible]		
					2	10.89	10.89			
					7	1.452	0.726			
					8	0.726	0.605	0.484		
					10	0.847	0.605			
					9	1.694				
3/20-07	Name Bhogi Mandal Address Kanakpur Lohna Road Darbhanga Occupation Service	Age 57 years Sex Male Weight 38.5 kg Height 1.52 m Complexion Dark	Dust and pollen allergy, cough, breathing trouble, loss of appetite, intermittent fever.	Since 10 years	6	1.452	0.968			
					4	0.484	0.726			
					1	1.21	1.21	1.21	1.21	1.21
					5	1.21	1.21	1.21	1.21	1.21
					3	4.84	[illegible]	3.63	2.42	
					2	21.78				
					7	1.694	0.605			
					8	1.21	0.605			
					10					
					9	1.694				
4/69-07	Name Ugra-narayan Jha Address Rariyari	Age 60 years Sex Male Weight 55 kg Height 1.8 m Complexion	Breathing trouble (due to asthma), high blood pressure, loss of appetite, distur-	Since 18 years	6	1.21	0.605	0.605	0.605	1.21
					4	0.363	0.484			
					1	1.21	1.21	1.21	1.21	
					5	1.21	1.21	1.21	1.21	

Table 2. Continued.

Serial No./ reference No.	Patient	Etiology	Symptoms	Duration of Disease	NOCC No	Doses of NOCC (g) I	II	III	IV	V
	Occupation		itching sensation		3	4.84	3.63	3.63	2,42	
	Service		in whole body,		2	10.89	10.89			
	(Headmaster)		abnormal pul-				–			
			pitation.		7	0.968	0,605	0.726		
					8	1,21	0,605	0.484		
					10					
					9	1.452				
					6 (Special)	0.847				
5/83-06	Name Kedar	Age 38 years	Sneezing, dust	Since	6	1.21	0.968			
	Pd. Sinha	Sex Male	allergy, breath-	5	4	0.968				
		Weight 55 kg	lessness (due to	years	1	1.21	1.21	1.21	1.21	1.21
	Address	Height 1.73 m	asthma)		5	1.21	1.21	1.21	1,21	1.21
	Mishratola	Complexion								
	Darbhanga	Medium								
	Occupation				3	7.26	7,26			
	Service				2	9.075	5,445	3.63	3.63	
	(Reader in									
	Zoology)									
					7	1.809	0.484	0.484		
					8	0.968	0,484			
					10					
					9					
6/109-06	Name	Age 25 years	Pollen allergy,	Since	6	1.452	0.726			
	Chandan	Sex Male	breathlessness	7	4					
	Choudhary	Weight 60 kg	(due to asthma)	years	1	0.968	0,968	0.968	0.968	0.968
	Address	Height 1.78 m			5	1.21	1.21	1.21	1,21	
	Nargawha	Complexion								
	Darbhanga	White			3	4.84	4,84	4.84		
	Occupation				2	10.89	10.89			
	Study				7	1 21	0.968			
	(MBBS				8	0.847	0.968			
	student)				10					
					9	1.694				

after administering first dose. Second dose also exhibited marvellous effect over joint pain, abnormal heart pulpitation and semen discharge along with urine. Slight relief in breathing trouble was also reported. After third dose patient reported normal appetite and sound sleep. However, no remarkable change was observed in breathlessness. Fourth dose controlled breathlessness, loss in concentration and confidence. The facial expression also improved a lot.

This patient also exhibited improvement in neutrophils, lymphocytes, eosinophils and monocytes, RBC, leucocytes, hemoglobin and ESR

Table 3. Differential counts of leucocytes of the patients before (a) and after 15 days of dose regulation (b). SE=1.94; CD=7.17.

Serial No./ Ref. No. of the patients	Neutrophils (%)		Lymphocytes (%)		Basophils (%)		Eosinophils (%)		Monocytes (%)	
	(a)	(b)	(a)	(b)	(a)	(b)	(a)	(b)	(a)	(b)
1/29-06	54	66	25	22	.6	.6	20	5	3	7
2/39-07	55	68	26	23	1	.5	8	4	4	6
3/20-07	66	66	24	24	1	1	25	6	13	8
4/60-07	65	65	22	23	1	1	14	5	13	12
5/83-06	60	67	25	22	1	1	22	4.5	7	7
6/109-06	67	67	20	20	.7	.7	5	3.5	10	10

(Tables 3 and 4). However, differences in lymphocytes, eosinophils, monocytes were found statistically insignificant.

Patient No. 3/20-07 reported no effect of the first dose but after administering second dose he repotted hot sensation in the internal parts of the body within 24 hours. Thereafter appetite increased and cough was gradually controlled. The third dose brought slight relief in breathing trouble but internal hot sensation persisted which turned into cold sensation after fourth dose. Fourth dose also reduced breathing trouble which was completely controlled after administering fifth dose. Symptoms of allergy also vanished. Weight increased by 2 kg.

Blood parameters of the patient No. 3/20-07 revealed improvement in basophils, eosinophils, monocytes, RBC, leucocytes, hemoglobin and ESR. The differences of basophils and monocytes were found statistically insignificant.

Patient No. 4/60-07 also reported no relief after administering first dose. The second dose aggravated breathing trouble and heart-pulpitation for 24 hours which gradually turned into more relief. Itching was reported. The third dose brought appetite normal. Sound sleep was also restorted. Relief in breathing trouble and itching was observed after this dose. Blood pressure was found normal. Fourth dose henceforth controlled breathing trouble normal. However, respiratory tract became quite clear after fifth dose. Gastritis was also controlled.

After six months the patient again reported breathlessness which was controlled after increasing one dose. The blood parameters of this patient exhibited improvement in eosinophils, monocytes and hemoglobin, but the differences of monocytes and hemoglobin were not statistically significant.

In patient No. 5/83—06, hot sensation in the inner parts of the body radiating to arms was reported after administering second dose. Sneezing, cough and allergy were controlled after third dose. Slight relief in breathing trouble was also reported. However, itching started in the whole body which was relived after removing $NOCC_4$. Fourth dose brought marked improvements in breathing trouble. Dose fifth induced more relief in breathlessness. $NOCC_9$ was yet to be administered.

Blood parameters of this patient exhibited improvements in neutrophils, lymphocytes, eosinophils, RBC, leucocytes, hemoglobin, and ESR. However, difference of neutrophils and lymphocytes were not statistically significant.

Patient No. 6/109-06 reported relief from pollen allergy after first dose. However, allergy completely vanished after administering second dose. Third dose covered marked relief in breathlessness except during athelatics. The breathlessness was controlled fully after fifth dose.

Table 4. Total counts of blood of the patients before (a) and after 15 days of dose regulation (b). SE= 0.276; CD=1.268; SE=0.3; CD=1.378.

Serial No./Ref. No. of the patients	Red Blood Corpuscles (RBC) (a)	Red Blood Corpuscles (RBC) (b)	Leucocytes (a)	Leucocytes (b)
1/29-06	4900,000	5000,500	9,200	8,000
2/39-07	4850,000	5100,000	12,000	7,000
3/20-07	4500,000	5000,000	14,000	9,000
4/59-07	5000,000	5.00,000	8,000	8,000
5/53-06	5200,000	54.0,000	11,000	10,000
6/109-06	5000,000	5000,000	10,500	10,000

Serial No./Ref. No. of the patients	Hemoglobin (g) (a)	Hemoglobin (g) (b)	Erythrocyte sedimentation rate (ESR) (mm) (a)	Erythrocyte sedimentation rate (ESR) (mm) (b)
1/29-06	11	13.5	13	11
2/39-07	10	14.5	15	12
3/20-07	8	13.0	20	14
4/59-07	14	15.0	12	12
5/53-06	14	15.5	12	10
6/109-06	15	15.0	12	12

Blood parameters exhibited improvements in eosinophils, and leucocytes while differences in eosinophils counts was found insignificant statistically.

It is by now crystral clear that NOCCs have certain effects on human diseases. The doses and combinations have specefic role. This opens a new vista in modern life science since, the mode of application is quite different from all other pathies. The action of NOCCs by simple tiding over body, recalls the recent findings of astrophysics regarding cosmic rays (1).

Cosmic Rays are high energy particles coming from sun, stars and galactic nebulae. They are of shorter wave-length. Most of them are in the range of 1 to 100000 Bev. It is such high energy particle which hits earth's atmosphere with 100 million Bev producing a gigantic shower containing millions of particles (1).

Naturally, we are surrounded by these high enery particles which must be interacting at least with our body physiology. As the cosmic rays coming from remote sources, they must be interacting among themselves and then with large chemical surfaces (planets). It is hoped that the cosmic rays are of different types depending upon the parts of the sky (360°). On the basis of effects (hot and cold) the sky has been divided into 12 parts (30° each) starting from central point of the equator. Thus, their interactions, normal and abnormal combinations and traits which interact with the body physiology produce normal and abnormal effects. Their cold effects, besides hot of varying degree and nature have been observed while interacting with the body physiology. The abnormal combinations of hot and cold induce abnormal action to body physiology leading to malfunctioning of the organs for particular biochmical functions appearing as diseases. The different sorts of allergy appear to be consequences of such abnormal combinations of hot and cold effects of the rays. These rays have been found to be giving three types of effects through energy support and heat cold effect (thermodynamic level) : Occasionally the rays at higher intensity for longer duration cause mutation ; with abnormal combination they cause abnormality to body-physiology ; and such abnormality at longer duration causes malfunctioning and damage to the organs.

This is similar to the action of electro-magnetic waves, as X-ray at low dose gives internal photograph but a particular high dose for a specified period causes mutation. Besides, hot and cold rays in adequate power and good angles induce beneficial effects. Similar sorts of actions have been displayed on algae (2, 3).

The combination of NOCCs gives soothing effect and ultimately recovery from ailments after administering required doses. The combinations refer to the combinations of hot and cold causing NOCCs may be due to regulation rays. They appear to set right the al combinations, powers and angles of the to the normal requirement of the body, NOCCs may be acting as 'rays alter (RADs). Hot and cold sensations in body and arms during regulation be due to changes induced at thermodynamic level on body physiological status, RAD be acting so due to their specific cond which have been further confirmed by imitations of such RADs to the body. Imi failed to give an effect. Aggravation in a reported by some of the patients dur lation may indicate that angles of rays portant. Some of the angles of the ray wrong for the body physiology whic be absorbed in greater quantity before ion through RAD aggravating the ailm hus, rays of wrong angles take some tim altered which is evidenced through agg turning into more soothing effects wit rt span of time. The rays might be acti genetic material, and biochemical functi patient provide evidence of changes about in blood characteristics.

smic rays appear to have great impac man being. They are of various types, ately hot and cold perhaps due to interactions, powers and angles. They maint in body physiological status under which organs function. They might be supporting as source of energy from outside to gene regulation. However, their characteristics are changing perhaps due to revolution and rotation of large chemical surfaces (planets) and causing individual differences due to their action at genetic level.

In allergy, abnormal combinations of rays appear effective since every good cold effect interact with bad hot rays causing malfunctioning of the lung, heart, bowel and blood. The medical conception dealt previously appears manifestations of such primary effects. RADs regulate primary effects giving radical cure to such patients which medical science fails to induce but only imparts temporary relief.

References

1. Rossi Bruno 1953. Where do cosmic rays come from I. Scientific American 189 64—70.
2. Srivastava P. and J. Nizam. 1988. Asterococcus, under high altitude cosmic radiations. Curr. Sci. 57 : 857—858.
3. Srivastava P. and J. Nizam. 1984. Effect of high attitude cosmic radiations on certain members of chlorococcales. Phykos 23 : 236—245.

एस्ट्रोजेनेटिक्स

रत्न-चिकित्सा की एक अनूठी पद्धति

● **अभिजित चटर्जी**

"एक हजार के लगभग ऐसे रोग हैं, जिनके बारे में आधुनिक चिकित्सा विज्ञान अनजान है और अब तक जिन रोगों के बारे में उसके पास जानकारी है, वह भी प्राथमिक नहीं, बल्कि माध्यमिक है। प्राथमिक जानकारी यह है कि रोगों का मूल कारण आकाशीय पिंडों से आनेवाली किरणें और उनका प्रभाव है।"

यह मान्यता है दरभंगा के प्रो. डॉ. छत्रनाथ झा की, जिन्होंने इस दिशा में अग्रणी और उल्लेखनीय कार्य किया है और हर प्रकार के हजारों रोगियों को स्वास्थ्य लाभ कराया है।

वैज्ञानिक आधार

व्यक्ति जब जन्म लेता है, उस क्षण व जन्म-स्थान पर जो ग्रह जिस स्थिति में होता है, वह उसको जीवन पर्यंत प्रभावित करता रहता है—ज्योतिष शास्त्रीय यह मान्यता आधुनिक विज्ञान की कसौटी पर खरी उतर चुकी है। अंतरिक्ष अनंत, और असीम है और अभी तक वह रहस्यमय बना हुआ है। उसमें असंख्य ब्रह्मांड (गैलेक्सियां) हैं। हमारी पृथ्वी की गैलेक्सी में कई सौर मंडल, अनगिनत तारे और नीहारिकाओं के जगमगाते पटल हैं, जो स्वयं को गुरुत्वाकर्षण परिधि द्वारा प्रायः अन्य सौर मंडल में पृथक रखते हुए अपना स्वतंत्र अस्तित्व स्थापित किये हुए हैं। इन समस्त आकाशीय पिंडों से इनकी किरणें पृथ्वी पर आती हैं। क्योंकि, ये आकाशीय पिंड पृथ्वी से करोड़ों मील दूर हैं, इसलिए इन किरणों को करोड़ों मील की दूरी तय करनी पड़ती है। इस दूरी को तय करने के बीच कुछ किरणें ओजोन गैस सतह द्वारा सोख ली जाती हैं, कुछ पृथ्वी के गुरुत्वाकर्षण बल के कारण वापस यानी परावर्तित हो जाती हैं। इस प्रकार कुछ किरणें ही पृथ्वी के वातावरण में प्रवेश कर पाती हैं। इन किरणों का प्रत्येक कण पृथ्वी के वातावरण

रत्न-चिकित्सा एक प्राचीन पद्धति है और वैज्ञानिक भी। इसी पद्धति पर गहन अध्ययन और शोध कर एक अनूठी चिकित्सा पद्धति विकसित की है दरभंगा के डॉ. छत्रनाथ झा ने। वह निशुल्क चिकित्सा कर हजारों असाध्य रोगियों को लाभ पहुंचा चुके हैं।

डॉ. झा दरभंगा के च. मि. वि. महाविद्यालय के वनस्पति-शास्त्र विभाग के आचार्य हैं । सन १९८३ की एक शाम आकाश को ताकते हुए उन्होंने ब्रह्मांड (गैलेक्सीज) के प्रति आकर्षण अनुभव किया और एस्ट्रोफिजिक्स के अध्ययन की ओर उन्मुख हुए । वर्षों के अध्ययन-मनन व शोधकार्य करते हुए उन्होंने रत्नों द्वारा रोगों की उपचार विधि विकसित की और उसे नाम दिया—एस्ट्रोजेनेटिक्स । और फिर उन्होंने विधिवत 'एस्ट्रोजेनेटिक्स सोसाइटी' का गठन किया । डॉ. झा के साथ उनके अनेक सहयोगी तथा अनेक युवा छात्र-छात्राएं इस सोसाइटी द्वारा रोग-मुक्ति के अभियान में आ जुटे । अब तक हजारों उन रोगियों को भी रोग-मुक्त करने में सफल रहे हैं, जिनको बड़े से बड़े डॉक्टर जवाब दे चुके थे । यह संस्थान रोगों का उपचार निशुल्क करता है । रत्न-उप-रत्नों का व्यय रोगी द्वारा वहन करना होता है । यह संस्थान एस्ट्रोजेनेटिक्स की उपचार पद्धति के प्रशिक्षण का पाठ्य-क्रम भी चला रहा है ।

डॉ. झा ने अपनी इस चिकित्सा-पद्धति की सूचना विश्व की प्रमुख वैज्ञानिक संस्थाओं को दी हुई है । जापान की साइंस काउंसिल ने जापान में इस विषय पर शोध के लिए एक विशेष समिति का गठन किया है । नयी दिल्ली स्थित आल इंडिया मैडीकल इंस्टीट्यूट ऑफ मैडीकल साइंस ने डॉ. झा की शोध में रुचि दिखलाते हुए उन्हें आमंत्रित किया है कि वह, उन्होंने जिन रोगियों को स्वस्थ किया है, उन्हें पूरे पिछले रेकॉर्ड के साथ लायें, ताकि उनका परीक्षण किया जा सके । जैनेटिक्स विभाग द्वारा प्रमाणित होने पर विश्व स्वास्थ्य संगठन तथा जापान फाउंडेशन और इंडियन काउंसिल फॉर मैडीकल रिसर्च उन्हें मान्यता प्रदान करने का आश्वासन दे चुका है ।

के साथ तीव्र रूप से प्रतिक्रिया कर किरणों की वर्षा करता है , इस से किरणें पृथ्वी के कण-कण में व्याप्त हो जाती हैं ।

कैसे डालती हैं प्रभाव

किरणों को करोड़ों मील की अपनी यात्रा के दौरान विभिन्न ग्रहों से होकर गुजरना पड़ता है, उनसे टकराकर उन्हें मार्ग बदलना पड़ता है । इसके कारण किरणों के तापक्रम तथा गुणों में भी परिवर्तन आ जाता है । प्रत्येक ग्रह अपने निश्चित पथ पर चलनेवाला तथा विशेष रासायनिक पदार्थों का सृजित पिंड है, जिसमें विभिन्न शीतोष्ण गुण भी होते हैं । यात्रा-क्रम में किरणें सबसे पहले बारह प्रकार से अंत-प्रतिक्रिया करती हैं । प्रतिक्रियाओं के क्रम में १२×१२×२७×२७×९=८५,०३,०५६ प्रकार की किरणें प्रत्येक मानव पर भिन्न कोणों तथा शक्ति से पड़ती हैं और उसके जीवन का संचालन करती हैं । ज्योतिष-शास्त्र के अनुसार जो १२ राशियां और २७ नक्षत्र मानव-जीवन का संचालन करते हैं उसका आधार भी यही है । आधुनिक शरीर विज्ञान भी इसी तथ्य को स्वीकार करता है और इसे अपनी भाषा में इस प्रकार कहता है कि मस्तिष्क क्रीनियल स्नायु के १२ जोड़े और २७ मुख्य स्पाइनक स्नायु द्वारा शारीरिक अंग-संचालन प्रक्रिया स्थापित करता है, जिसमें पूर्ण अंग-क्षमता भी परिलक्षित होती है । अर्थात प्रथम क्रीनियल स्नायु जिस अंग से जुड़े होते हैं, बारह राशियों में प्रथम राशि—मेष भी उसी अंग का परिज्ञान स्थापित किये हुए है और इसी प्रकार अन्य भी । इन मौलिक गुणों में समानता के कारण ज्योतिष शास्त्र में अनुपलब्ध पूर्ण-अंग-विचार जीवन-विज्ञान की सहायता से

उपलब्ध हो गये हैं । इसी वैज्ञानिक आधार पर डॉ. छत्रनाथ झा ने अपनी रत्न चिकित्सा प्रणाली को विकसित किया है ।

डॉ. छत्रनाथ झा का कहना है कि ८५,०३,०५६ प्रकार की किरणें प्रत्येक मानव पर विभिन्न कोणों तथा शक्ति से पड़कर उसका संचालन करती हैं । इन किरणों में दूषित या विध्वंसकारी किरणें तो होती ही हैं, साथ ही कुछ लाभकारी किरणें भी जन्म-स्थान के अक्षांश व देशांतर व समय के कारण खास कोण से आती हुई विध्वंसकारी हो सकती हैं । दूषित या विध्वंसकारी किरणें जब मानव शरीर के किसी अंग का संचालन खास समय में, खास कोण से करने लगती हैं, तो उस समय उस अंग में रोग विशेष पैदा हो जाता है । रोग पैदा होने की स्थिति खासकर तब बनती है, जबकि अन्य लाभकारी किरणों की संचालन-शक्ति के क्षीण हो जाने से शरीर की प्रतिरोधक-प्रक्रिया समाप्त हो जाती है ।

उपचार पद्धति

जिन दूषित किरणों के प्रभाव से रोग उत्पन्न होते हैं, उनका उपचार रत्न और उप-रत्नों द्वारा किया जा सकता है । इसके लिए रत्न और उप-रत्नों का उचित योग एवं मात्रा में उपयोग किया जाता है । डॉ. झा और उनके सहयोगियों ने जो अध्ययन-मनन और शोध किये हैं, उनका निष्कर्ष है कि मूंगा (इटालियन), गार्नेट, टोपाज, मोती, लाजवर्त्त, गोमेद, टाइगर्स, स्फटिक, ओनेक्स, ए. डी., पुखराज, पन्ना, हीरा, फिरोजा आदि का विभिन्न योग एवं मात्रा विभिन्न रोगों के निवारण के लिए अत्यंत लाभकारी है । अधिकांश रोगों का निवारण उप-रत्न से भी संभव हो जाता है, लेकिन इसके प्रयोग की प्रक्रिया जटिल है । इन रत्नों-उप-रत्नों के उचित योग व मात्रा रोगों से बचाव यानी रक्षा-कवच का कार्य करता है

रत्न : किस तरह

रत्न और उप-रत्न एक विशेष प्रकार के केमिकल कंपाउंड होते हैं । कास्मिक किरणें विशेषकर बारह कास्मिक किरणें तीस-तीस डिग्री के आकाश पिंड से आती हैं । किरणों में अंत-प्रतिक्रिया फिर से ग्रहों से प्रतिक्रिया होती है । इस प्रकार किरणें अनंत किस्मों की हो जाती हैं, लेकिन जब खास प्रकार की किरण खास भू-खंड के खास रसायन से लगातार प्रतिक्रिया करने लगती हैं, तो वह रासायनिक पदार्थ उस किरण का अवशोषण गुण प्राप्त कर हजारों वर्ष में रत्न व उप-रत्न बन जाता है । रत्न व उप-रत्नों के सृजन की विधि प्राकृतिक एवं धीमी होती है, इसलिए इनकी मात्रा कम ही पायी जाती है । इसलिए ये 'कीमती पत्थर' होते हैं । उप-रत्न कम ही समय में किरणों के प्रभाव से बना हुआ, कुछ भिन्न रासायनिक पदार्थ, कम क्षमतावाले किरण ग्राही के रूप में विकसित पत्थर होते हैं ।

दूषित कोण से आती हुई किरणें व्यक्ति पर दुष्प्रभाव डालती हैं इन दुष्प्रभावों का परिमार्जन किरण-ग्राही पदार्थ रत्न या उप-रत्न की उचित मात्रा एवं योग से संभव हो जाता है ।

डॉ. झा का पता है—*संगीत सदन, कठहल बाड़ी, दरभंगा (बिहार)*

—१०, बीरबल रोड,
जंगपुरा एक्सटेंशन
नयी दिल्ली-११००१४

Miracle cure for liver cirrhosis!

The Times of India News Service
DARBHANGA January 15:

IN what seemed to be a miracle, a patient, suffering from liver cirrhosis, who had been refused admission at the Delhi-based All India Institute of Medical Sciences (AIIMS) because of his serious condition, was cured by Dr C N Jha, an astro-genetic scientist and life science teacher at the C M Science College, here.

Dr Jha had prescribed the patient to wear some stones for a stipulated period of time. The patient responded to the treatment, and within three months, he was declared healthy by the same AIIMS after different pathological tests.

This, incidentally, is not the only case of its kind, for this researcher has created a sensation in the field of medical science by curing patients with terminal diseases. He is now engaged in conducting an intensive study on astro-genetic science and its use for the treatment of several killer diseases.

"It was not a freak chance but a miracle of empirical science", says Dr Jha. He has treated more than 30 terminally ill patients since then and claims to have succeeded in about 80 per cent of the cases. "Failures are mainly because of discontinuation of the therapy and incorrect astrological data", points out Dr Jha.

"Astro-genetics", says Dr Jha, "explores into a new area of therapeutic system for the treatment of human diseases". According to him, the malfunctioning of the biological system of an individual is largely due to resultant conditioning caused by the constant exposure of human body to cosmic rays. In this respect, astro-genetics is based on the same principles on which astro-physics is founded.

Dr Jha calculates the astrological data of an individual's planetary position at a particular time to treat the disease with the aid of 'naturally occurring chemical compounds' (NOCC) like stones and gems. He maintains that stones and gems have so far been prescribed to bring good luck or prevent mishaps for an individual. Making a scientific study of astrological premises and experimenting them with NOCC (stones), he has conclusively figured out that human diseases can be treated with an astro-genetic system of therapy. For this, he prescribes different stones with all possible manipulations, dosing, duration-variation and anatomical variation.

Dr Jha claims to have cured patients suffering from different incurable diseases like paralysis, mental retardation, spondylitis, myopia, cancer, intestinal problems and other chronic diseases. He, however, admits the limitation of the system in the treatment of traumatic, epidemic and immune-deficiency syndrome cases.

Stone researcher's claim to cure terminal ailments

The Times of India News Service

Darbhanga, January 15: In what seemed to be a miracle, a patient suffering from liver cirrhosis, who had been refused admission at the Delhi-based All India Institute of Medical Sciences (AIIMS) because of his serious condition, was cured by Dr C N Jha, an astro-genetic scientist and life science teacher at the C M Science College, here.

Dr Jha had prescribed the patient to wear some stones for a stipulated period of time. The patient responded to the treatment, and within three months, he was declared healthy by the same AIIMS after different pathological tests.

This, incidentally, is not the only case of its kind, for this researcher has created a sensation in the field of medical science by curing patients with terminal diseases. He is now engaged in conducting an intensive study on astro-genetic science and its use for the treatment of several killer diseases.

"It was not a freak chance but a miracle of empirical science", says Dr Jha. He has treated more than 30 terminally ill patients since then and claims to have succeeded in about 80 per cent of the cases. "Failures are mainly because of discontinuation of the therapy and incorrect astrological data", points out Dr Jha.

"Astro-genetics", says Dr Jha, "explores into a new area of therapeutic system for the treatment of human diseases". According to him, the malfunctioning of the biological system of an individual is largely due to resultant conditioning caused by the constant exposure of human body to cosmic rays. In this respect, astro-genetics is based on the same principles on which astro-physics is founded.

Dr Jha calculates the astrological data of an individual's planetary position at a particular time to treat the disease with the aid of 'naturally occuring chemical compounds' (NOCC) like stones and gems. He maintains that stones and gems have so far been prescribed to bring good luck or prevent mishaps for an individual. Making a scientific study of astrological premises and experimenting them with NOCC (stones), he has conclusively figured out that human diseases can be treated with an astro-genetic system of therapy. For this, he prescribes different stones with all possible manipulations, dosing, duration-variation and anatomical variation.

Dr Jha claims to have cured patients suffering from different incurable diseases like paralysis, mental retardation, spondylitis, myopia, cancer, intestinal problems and other chronic diseases. He, however, admits the limitation of the system in the treatment of traumatic, epidemic and immune-deficiency syndrome cases.

The Science Council of Japan and the genetic unit of the AIIMS have shown their interest in Dr Jha's new-found branch of medication. The AIIMS, in fact, has already invited Dr Jha for further discussions with the experts of the concerned fields there.

ATE THE INDIAN NATION, FRIDAY, PATNA, JULY 14, 1995 | 3

Astrogenetics cure for terminal patients

IN Correspondent

DARBHANGA, July 13: 'No astrological occultism but an empirical science', says Dr. C.N. Jha, a life-science teacher of C.M. Science College about his exclusive innovation in the therapeutic system of Astrogenetics. Surrounded by patients, mostly of terminal diseases like liver cirrhosis, abdominal cancer, paralysis and jaundice at an advanced stage, Dr. Jha explains what exactly Astrogenetics stood for.

Popularly known as 'stone-therapy' because stones of various chemical substances are modulated on the patient's body in accordance with the patient's astrological data. Modulation is staggered in duration, doses and combines, and the patient is held in strict observation to monitor the effect which, he said, "is perceptible within hours-manifestly in aggravation, stabilisation or pain-relief'. This is, according to Dr. Jha, response-syndrome that leads him encouragingly to further modulation.

Treatment continues for one to three months' duration depending on the seriousness of disease. This is in no case, he asserts, groping in the dark, rather if the astrological data are deciphered correctly, prescription of stones in combines is objectively verifiable. And the effect is that way predetermined. 'So there is nothing occult about it', he said.

Discussing the types of diseases he successfully treated and cured. Dr. Jha says diseases caused by genetic malformation respond tremendously to stone-therapy, though other diseases, too, are efficaciously treated in this system.

He claimed to have cured some of the lost cases particularly in cancer and liver cirrhosis which had eventually been refused by IIMS and other prestigious medical institutes.

He has with him, cases of asthma, allergy, acute gastritis, diabetes, spondalitis, gout etc. besides many other ailments. Percentage of successes varies to 70 to 90 per cent, depending on accuracy of astrological details, continuity of therapy and genuineness of stones. He, however did not claim to have treated any AIDS case, though he hopes of the effectiveness of astrogenetic system of medicine also on patients testing HIV positive. With regards to traumatic cases particularly in pestilence, his therapy might not be advisable.

The theory of the treatment is guided not by the type of the disease but by the type of rays, emitted from the cosmos that an individual absorbs in the system. Malignant rays that an individual receives due to the influence of a particular planet on the system are negated or shaded while the benign rays are to be reinforced through stones as catalysts.

Constant exposure of an individual to some malignant rays cause various genetic malformation to develop in the human system. Thus, stones start repairing the malformation immediately the rays are prevented from influencing the particular human system.

So, astrogenetics is also a preventive medicine. On each individual's biological system of existence, there is the concentration of particular cosmic rays at a particular point or period of time, depending on an individual's astrological constitution. Planetary position at the time of birth predominates the incoming physical variables, Dr. Jha says. Managing these rays according to individual requirement is the crux of health-management strategy in astrogenetics.

In Dr. Jha's efforts to disseminate information on complementary health-care and his exploration in a new branch of healing-system. All India Institute of Medical Sciences, Indian Council o[illegible]al Research and Science Council of Japan took active interest in [illegible]netic system of therapy. Dr. Jiro Kodo of Science Council of Japan recommended Dr. Jha's-astrogenetic theory to Dr. Ekwo Takenchi president of Research Liaison Committee for Comments.

Dr. (Professor) I.C. Verma of IIMS-Pediatrics called for complete information about 30 patients who were not cured of their ailments at IIMS and treated successfully through Dr. Jha's stone-therapy.

In Dr. Jha's views, through extensive research and development, astrogenetic can be developed into a complementary branch of genetic engineering, and then even those ailments which defy effective genetic correction may be treated much more effectively.

For that, necessary infrastructure and wherewithal are required. Nevertheless, Dr. C.N. Jha keeps on mitigating sufferings of humanity to the extent possible for him.

लीवर सूत्रण रोगी चमत्कारिक रत्न उपचार से चंगा

गंभीर रोगी की भरती लेने वाले आयुर्विज्ञान संस्थान ने डा० झा के इलाज के बाद स्वस्थ घोषित किया

दरभंगा २८ जनवरी। इसे चमत्कार ही माना जाएगा कि लीवर सिरोसिस (सूत्रण रोग) से पीड़ित जिस मरीज को उसकी चिंताजनकदशा देख दिल्ली के अखिल भारतीय आयुर्विज्ञान संस्थान ने भर्ती लेने से इंकार कर दिया था उसे यहां सी एम साइंस कालेज के अनुवांशिकी रत्न ज्योतिष विज्ञानी एवं जननिक विज्ञान अध्यापक डा० सी एन झा ने चंगा कर दिया।

डा० झा ने मरीज को एक निश्चित अवधि तक कुछ रत्न धारण करने को कहा। मरीज ने डा० झा के बताये रत्न धारण किये और उन नगीनों ने अपना प्रभाव दिखाया और तीन मास बाद अखिल भारत आयुर्विज्ञान संस्थान ने ही विभिन्न पैथोलाजी परीक्षणों के बाद मरीज को एक दम स्वस्थ घोषित किया। रत्नों ,कीमती पत्थरों द्वारा असाध्य रोग का सटीक उपचार का यह अपने ढंग का अकेला केस नहीं था। इस ज्योतिष विज्ञानी –शोधकर्ता ने बीमारी के अंतिम चरण में पुहंच चुके मरीजों को स्वस्थ करके चिकित्सा विज्ञान के क्षेत्र में सनसनी फैला दी है। डा० झा इन दिनों अनेक जानलेवा बीमारियों के रत्न ज्योतिष द्वारा उपचार का गहन अध्ययन कर रहे है।

डा० झा कहते हैं कि रत्न ज्योतिष द्वारा बीमारी के नाजुक मोड़ पर पहुंचे मरीजों का चंगा होना कोई तुक्का नहीं हैं। बल्कि यह अनुभव जन्य विज्ञान का करिश्मा हैं। डा० झा तबसे ३० से अधिक ऐसे मरीजों को चंगा कर चुके हैं जिनकी बीमारी प्रायः अंतिम पड़ाव पर पहुंच चुकी थी। डा० झा का दावा है किइन केसों में उन्हें करीब ८० प्रतिशत सफलता प्राप्त हुई है। २०प्रतिशत असफलता मरीज द्वारा रोगोपचार में नागा तथा सटीक ज्योतिष गणना के अभाव के चलते हुई।

डा० झा कहते हैं कि जननिक ज्योतिष ने मानवीय बीमारियों के इलाज हेतु रोगोपचार पध्दति में एक नये क्षेत्र को तलाशा है। उनके अनुसार ब्रह्मांड-किरणों का मानव शरीर पर लगातार अनावृत्त पड़ना तथा उससे उत्पन्न विकिरण बीमारियों के कारण होते है। जननिक रत्न ज्योतिष उन्हीं सिद्धान्तों पर आधारित है जिन पर खगोल भौतिकी की आधार शिला रखी है। सटीक ज्योतिष गणना निश्चित बीमारी में रत्न विशेष धारण किये जाने पर बीमार का चंगा होना शत प्रतिशत निश्चित है।

"आँदीव" दैनिक 21 जनवरी १९८५

उ० प्र० दैनिक

आर्यावर्त, पटना, बृहस्पतिवार, १३ जून, १९९...

रोगों की चिकित्सा में रत्नों का चमत्कारी प्रभाव

आज के इस वैज्ञानिक एवं कम्प्यूटर युग में रत्नों का प्रयोग फैशन एवं सौंदर्य प्रदर्शन हेतु किया जा रहा है लेकिन गंभीर रोगों की चिकित्सा के दृष्टिकोण से भी इन रत्नों का एक अपना अलग महत्व है। वर्तमान समय में एलोपैथिक रोग चिकित्सा प्रविधि चरमोत्कर्ष पर है। हृदय, आंख, किडनी तथा कई अन्य अंगों का अब सफल प्रत्यारोपण भी होने लगा है जो आधुनिक चिकित्सा जगत की महान उपलब्धि है। प्राचीन काल से ही ज्योतिष के आधार पर रोगों की जानकारी प्राप्त करना तथा इनके निदान हेतु अनुकूल परिणाम प्राप्त करना महर्षियों की महान उपलब्धि रही है और आज भी इस संदर्भ में सतत् अन्वेषण एवं भगिरथ प्रयत्न जारी है। भारतीय ज्योतिष ने अब कुण्डली के आधार पर रोगों की जानकारी देकर तथा उसका निदान अच्छी-खासी सफलता अर्जित की है। रोगों की उत्पति क्यों होती है–इस संदर्भ में ज्योतिष में कहा गया है कि कर्म से रोग की उत्पति होती है। कुण्डली में ग्रहों का निरीक्षण कर यह निष्कर्ष निकाला जा सकता है कि किन-किन ग्रहों से जातक को कौन-कौन सी बीमारियां होंगी। इस आधार पर सूक्ष्म अध्ययन कर जीवन में होनेवाले रोगों का पता लगाया जा सकता है। 'जातक परिजात' में कहा गया है कि 'अगर रोग है तो इसका कारण है और निदान भी है। ज्योतिष में रोग का कारण ... माना गया है ...

गोमेद तथा लहसुनिया। इसके अतिरिक्त अन्य रत्नों को उपरत्न की संज्ञा दी गयी है। समाज में इसी नौ रत्नों को ही रत्न के रूप में स्वीकार किया गया है। इनमें माणिक्य, मोती, हीरा, नीलम तथा पन्ना को महारत्न कहा गया है लेकिन पौराणिक मान्यता है कि 'जब समुद्र मंथन हुआ तो चौरासी रत्न प्राप्त हुए थे। इनकी रूप रेखा क्या थी इसका वर्णन नहीं मिलता, लेकिन आज के वैज्ञानिक नये-नये खनीजों को भी रत्नो का नामकरण करते जा रहे हैं फलस्वरूप आज रत्नों की संख्या सौ तक पहुंच गयी है। इसके अतिरिक्त 'जयपुर जौहरी एशोसिएशन' जिन रत्नों को मान्यता प्रदान किया है उन रत्नों की संख्या भी चौरासी ही है। जिन रत्नों को महारत्न की संज्ञा दी गयी है वह तत्काल अपना प्रभाव दिखाते हैं।

रोग, गठिया, जौंडिस जैसे असाध्य रोगों का इलाज किया जा चुका है। प्राकृतिक चिकित्सा की एक पद्धति वर्ण चिकित्सा (क्रोमोपैथी) का भी कहना है विभिन्न प्रकार के रोगों में व्यक्ति को निरोग करने के लिए रंगों में असाधारण शक्ति है। सुप्रसिद्ध चिकित्सा शास्त्री डा. इडगर मैसी का कहना है कि चिकित्सा का भविष्य विभिन्न रंगों की प्रकाश किरणों और ध्वनि तरंगों पर निर्भर करता है। मैंने स्वयं स्टोन थेरापी द्वारा कई गंभीर रोगों की चिकित्सा में चमत्कारी प्रभाव पाया है।

शोध एवं व्यवहारिक अनुभव के आधार पर निम्नलिखित रोगों के निदान हेतु रत्नों का प्रयोग किया जा सकता है–

माणिक्य–घबड़ाहट, बेचैनी, धड़कन, भय, भ्रम, वक्षपीड़ा, कमजोरी, यादाश्त की

प्रो० प्रेमचन्द्र पाण्डेय

प्रो॰ ...

नेत्र रोग।

हीरा–वीर्य दोष, धातु रोग, शीघ्र पतन प्रमेह, दौर्बल्य क्षीणता नपुंसकता, धर्मक्षय

नीलम–विषम ज्वर, मिर्गी, मस्तिष्क की कमजोरी, उन्माद, हिचकी, स्नायविक दर्द, गठिया, संधिवात, उदरशूल, वायु, बेहोशी, पक्षाघात, बवासीर, वात रोग, हर्निया गठिया।

गोमेद–वायु शूल, गैस्ट्रीक, चर्म रोग, कृमि रोग, बवासीर, फाइलेरिया, गर्मी, हैजा।

लहसुनिया–कफ, खांसी, बवासीर, पित्त दोष, वीर्य दोष, अंडकोश के रोग, विष, खून की कमी, चेचक।

रत्न धारण पूर्व आवश्यक सावधानी

(१) रत्न धारण पूर्व कुंडली का सूक्ष्म अवलोकन करना आवश्यक है और रोग से संबंधित ग्रह के शुभ प्रभाव तथा पीड़ित अवस्था में रोग से संबंधित ग्रह रत्न को धारण किया जाय।

(२) जिस ... को धारण करना है उसे ...

...रोगोंपर रत्नका औषधीय प्रभाव

...ग्रन्थोंके आधारपर रत्नोंका रोग निदान हेतु प्रयोगका संक्षिप्त विवरण प्रस्तुत कर रहीं हूं-

रत्नोंको औषधके रूपमें प्रयोगमें लाते समय रत्नोंके गुण एवं उसके रासायनिक स्वरूपको ध्यानमें रखकर उसके भस्म, पिष्टी, जल या तैल रूपका प्रयोग किया जाता है।

अब हम रत्नोंके क्रमशः औषधीय गुणोंको देखें।

माणिक्य- भावप्रकाशमें माणिक्यको काषाय और मधुर रस प्रधान बताया गया है। शास्त्रोंमें 'माणिक्यं दीपनं वृष्यं कफवातक्षयातिनुत्' कहकर शरीरमें उत्पन्न उष्णता, जल,

रत्न चिकित्सा

कफ, वात तथा क्षय रोगका नाशक 'माणिक्य' को कहा गया है। घबराहट, बेचैनी, धड़कन, भय, भ्रम, वक्षपीड़ा, कमजोरी, ब्लडप्रेशर, याददाश्तकी कमी, हृदयरोग, रक्तचाप, दुर्बलता, पेटदर्द एवं आंखोंकी बीमारियों आदिमें काफी लाभकारी इसे बताया गया है।

मोती- मोती कैल्शियमका सर्वोत्तम भरपाई करनेवाला रत्न है। आयुर्वेदके भावप्रकाशमें 'मौक्तिकं शीतलं वृष्यं चक्षुष्यं बलपुष्टिदम्' कहकर इसे शीतल, वीर्यवर्धक बल पुष्टिदम् एवं आंखोंके लिए उपयोगी तथा बलकारक बताया गया है। अचेतनता, मूर्छा, हिस्टीरिया, कैन्सर, एलर्जी, क्रोध, अनिद्रा, मानसिक विकार, मूत्ररोग, दन्तरोग, भय, भ्रम, दिमागकी बीमारियोंके साथ ही वायुविकार, खूनी बवासीर, मस्तिष्क ज्वर तथा पित्त सम्बन्धी गड़बड़ियों, श्वेत प्रदर एवं रक्त प्रदर आदिमें भी 'मोती' का उपचार प्रयोग किया जाता है।

मूंगा- रक्तकी अशुद्धता, कमी, शिथिलता, निस्तेजता, कैन्सर, रक्तदोष, बवासीर, अमाशय सम्बन्धी रोग, यक्ष्मा, कुष्ठ, नेत्ररोग, पित्तरोग, पीलिया, हृदयकी अत्यधिक धड़कन आदिके साथ ही गुलाब जलमें घिसकर गर्भवती स्त्रीके पेटपर लेप करनेसे गर्भस्राव नहीं होने देता है तथा त्वरित गर्भको भी रोक देनेका गुण मूंगामें विद्यमान है।

पन्ना- आयुर्वेदके ग्रन्थ 'रस समुच्चय' (रस रत्न समुच्चय) में पन्नाको श्रेष्ठ विषनाशक बताया गया है। श्वास, दमा, खांसी, अल्सर, रक्तचाप, टांसिल, अनिद्रा, सन्निपात, प्रमेह, मिचली आदिको नष्ट कर बुद्धि एवं वीर्यवर्धनकी अचूक शक्तिको रखता है।

> आधुनिक परिवेशमें रत्नोंका उपयोग और भी बढ़ गया है। समाजके १५ प्रतिशत लोग किसी न किसी रूपमें रत्न धारण किये अवश्य देखे जा सकते हैं। रत्नोंके उपयोग तथा धारणसे यह स्पष्ट हो गया है कि यह मानव मस्तिष्कको प्रभावित करता है। यही कारण है कि धीरे-धीरे इनका उपयोग ग्रहोंको मजबूत करने तथा रोगोंके उपचारके लिए किया जाता है।

पुखराज- यह स्वयंमें एक महौषधि होते हुए मुख सम्बन्धी रोग, गुदा, मूत्राशय, यकृत, तिल्ली, रक्तचाप, कण्ठरोग, गुप्तेन्द्रिय रोग, आन्त्रशोथ, प्रदर, कैन्सर, हड्डियोंके दर्द, सिरदर्दको नष्ट करनेवाला होता है।

हीरा- रस रत्न समुच्चयमें वर्णन मिलता है कि अगर कोई रोगी मृत्युके करीब हो तो ऐसी अवस्थामें यदि हीरेकी भस्म उसे दे दिया जाय तो उसकी डूबती सांसें भी लौट आती हैं। वीर्यदोष, धातुरोग, शीघ्रपतन, प्रमेह, नपुंसकता, गर्भक्षय, मधुमेह, टी. बी., एनीमिया, अनुपतन, मोतियाबिन्द, हिस्टीरिया आदि रोगोंपर रामबाणके समान काम करता है। (सावधानी- हीरेको खानेसे मृत्युतक हो जाती है अतएव कोई भी औषधि बिना चिकित्सकके परामर्शके न ली जाय)।

नीलम- यह रत्न विषम ज्वर, मिरगी, मस्तिष्ककी कमजोरी, पागलपन, हिचकी, गठिया, सन्धिवात, पेटदर्द, बेहोशी, पक्षाघात, बवासीर, हर्निया आदि रोगोंपर उपयोग किया जाता है। कुष्ठ, टी. बी., खांसी तथा नेत्र विकारकी भी अचूक दवा इसे माना जाता है।

गोमेद- यह वायु शूल, चर्मरोग, कृमि रोग, बवासीर, मलेरिया, गर्मी, हैजा आदिमें विशेष लाभप्रद है। इसके अतिरिक्त गोमेदको बुद्धिवर्धक, कफनाशक तथा मस्तिष्ककी दुर्बलताको दूर करनेमें उपयोगी बताया जाता है।

लहसुनिया- इसका उपयोग मूलतः पित्तजनित रोगोंके लिए किया जाता है। कफ, खांसी, बवासीर, पित्तदोष, वीर्यदोष, अण्डकोशके रोग, खूनकी कमी, चेचक, नेत्ररोग हारक तथा बल-बुद्धि एवं वीर्यवर्धन में भी लहसुनियाका प्रयोग किया जाता है।

इसके अतिरिक्त आजवत मिरगीमें, दाना फिरंग गुर्देकी खराबीमें, मिर्गीलिया कुलपित्तीमें, अम्बरका प्रयोग सन्निपात रोगमें, ओपलका ज्वरमें, तामड़ाका पथरीमें, हकीकका बेहोशीमें, पुखराजका मुखके दुर्गन्धमें, हौलदिलका हृदयकी धड़कनमें, फिरोजाका प्रयोग आंखोंके सूजनमें भी किया जाता है।

प्रायः सभी रत्नोंका प्रयोग भस्म, पिष्ट, चूर्ण एवं जलके माध्यमसे भी किया जाता है। इन सभीका सेवन किसी योग्य चिकित्सक, कविराज या वैद्यकी देखरेख एवं परामर्शपर ही होना चाहिए, अन्यथा परिणाम घातक भी हो सकता है। हीरेकी पिष्टी कभी नहीं खानी चाहिए। भूलवश अगर हीरेका कोई कण पेटमें चला जाय तो दूधमें घी मिलाकर पीनेसे उल्टीके साथ हीरेका कण बाहर आ जाता है। हीरेका कण अगर पेटमें रह जाता है तो वह पेटकी अंतड़ियोंको क्षतिग्रस्त कर देता है जिससे मृत्युतक हो सकती है।

● पूनम दिनकर

शनिवार, २२ अगस्त १९९८ आज

आकाशीय किरणोंका प्रभाव मानव शरीरपर पड़ता है

95

दरभंगा, १५ दिसम्बर (आ. सं.)। पीड़ित मानवताकी सेवा एवं आकाशीय कौस्मिक किरणका मानवपर पड़ने वाले प्रभावके अध्ययनको मूल उद्देश्य बनाकर एस्ट्रोजेनेटिक सोसायटी पांचवें वर्षमें प्रवेश कर रही है। आधुनिक जैव विज्ञानजो पूर्ण रूपेण जीवोंपर आकाशीय किरणके प्रभावको नजरंदाज कर विभिन्न शोध कार्य सम्पन्न कर रहा है। उसके समक्ष यह चुनौती भरा प्रश्न है कि आकाशीय किरण जो पृथ्वीके निर्माणमें अहम भूमिका अदा की है, अब तक जैव विज्ञानके लिए अछूता क्यों है?

एस्ट्रोफिजिक्सके विद्वानोंने यह प्रमाणित किया है कि अंतरिक्ष किरणें सूर्य, तारे, [illegible] आकर पृथ्वीके गुरुत्वाकर्षण को पार कर प्रवेश करती है जो अत्यन्त ही शक्ति शाली है। यह विज्ञानके समक्ष एक सशक्त प्रश्न उठाता है कि इतनी शक्ति संपन्न किरणोंसे घिरी पृथ्वीपर का 'जीवमंडल' इसके प्रभावसे किस प्रकार बच सकता है? इस प्रश्नके निदान के लिए पांच वर्ष पूर्व स्थापित की गयी एस्ट्रोजेनेटिक सोसाइटी नामक यह संस्था अभी तक के शोधोंसे यह साबित कर चुका है कि आकाशीय किरणोंका प्रभाव मानवीय शरीरपर पड़ता है जिससे शरीरका आंतरिक वातावरण नियंत्रित होता है हजारों रोगियोंपर किए गए अबतकके शोधोंसे यह जाहिर हो चुका है कि रोगों एवं आंतरिक गुणोंका कारण आकाशीय किरणें हैं जिनका निराकरण रत्नों एवं उपरत्नों के द्वारा हो सकता है। यह जानकारी सोसायटीके निदेशक डा. सी. एन. झा. ने दी।

हिन्दुस्तान,
जनवरी - 3
1995

एस्ट्रोजेनिक सोसाइटी

नव वर्ष के उपलक्ष्य में, ऐस्ट्रोजेनेटिक पद्धति से इलाज कर रही ऐस्ट्रोजेनेटिक सोसायटी (दरभंगा) १ से १५ जनवरी के बीच आने वाले मरीजो के इलाज के लिए सामाजिक सहयोग से रत्नो व उपरत्नो की पहली खुराक मुफ्त देगी। स्थानीय कटहलबाड़ी मुहल्ला स्थित 'संगीत सदन' में आयोजित मानव कल्याण यज्ञ में सभी के सहयोग की अपेक्षा की गयी है।

भारत सरकार
आकाशवाणी
Government of India
ALL INDIA RADIO
वार्ता/लघु कथा शाखा
TALKS/SHORT STORY BRANCH
दरभंगा

आकाशवाणी कार्यक्रम-1
(देख 5-10-11/5-10-15 देखिए)
AIR-P-I
(See Paragraph 5-10-11/5-10-15)

"हिन्दी अनुभाग"

सेवा में,

संख्या
No.........

दिनांक
Dated..20-8-1991...

डा. ...., टी.ए.
.... विभाग, सी.एम. कालेज,
दरभंगा

इस पत्र के पीछे छपी शर्तों पर हम नीचे लिखे विषय, दिनांक और समय पर आपकी वार्ता/लघु कथा सहर्ष प्रसारित करेंगे। कृपया संलग्न पुष्टिपत्र पर हस्ताक्षर करके इसको लौटती डाक से भेज दीजिए। इसके लिए हम आपके आभारी होंगे।

We shall be pleased to broadcast your talk(s)/short story on the subject, date and time detailed below upon the conditions printed overleaf. We shall be obliged if you could kindly sing and return the attached confirmation sheet not later than ....

शीर्षक / Title: वार्ता: "रत्न चिकित्सा द्वारा रोगों का निदान एवं निवारण"

तारीख (तारीखें) / Date(s): प्र० तिथि 24-8-91 रिका० 23-8-91: 2 बजे

प्रसारण का समय / Time of Broadcast: रात 8:00 बजे

प्रसारण की अवधि / Duration: 9 मिनट

प्रसारण का स्थान / Place of Broadcast: आकाशवाणी, दरभंगा

फीस / Fee: 100/-(एक सौ रुपये मात्र)

अनुवर्ती प्रसारण शुल्क / Subsequent Broadcast fee ....

(इस पत्र के पीछे छपी शर्तों के खण्ड 4(क)/4(ख) एवं 4(ग) के अध्यधीन)
(Subject to Clause 4(a)/4(b) & 4(c) of conditions printed overleaf).

हमारा विशेष निवेदन यह है कि आप कृपया इस शर्त का पालन करके हमारी सहायता करें कि प्रसारण के लिए जो तारीख निश्चित की गई है उससे कम से कम दस दिन पहले वार्ता/लघु कथा की पांडुलिपि केन्द्र निदेशक के पास पहुँच जाए। इस शर्त का पालन न होने पर केन्द्र के सामान्य कार्यक्रम में गम्भीर अवरोध पैदा हो जाता है।

We would particularly ask you to assist us by complying with the condition that the manuscript of the talk/short story should be in the hands of the Station Director not less than ten days before the date fixed for the broadcast. The normal routine of the Station is seriously hampered if the condition is not observed.

स्टाम्प शुल्क सरकार द्वारा वहन किया जाएगा।
The Stamp Duty will be borne by the Government.

भवदीय
Yours Faithfully,

कृते केन्द्र निदेशक
for Station Director

भारत के राष्ट्रपति के लिए और उनकी ओर से
for and on behalf of the President of India.

टिप्पणी
लघु कथा के मामले में लेखक उसका शीर्षक यहां भरें।
In case of Short Story the title of the Short Story will be filled in by the author

Telegram— "AKASHVANI" [illegible]

Government of India

(See Paragraph 5-10-11,5-10-15)

ALL INDIA RADIO

TALKS/SHORT STORY BRANCH

* हिन्दी *

सेवा में,

डा० डॉ० सी० एन० झा, रीडर

... विभाग, सी०एम० साइंस कालेज,

दरभंगा

संख्या

No. ...

दिनांक

Dated 21-11-91 ...

महोदय/महोदया

Dear Sir/Madam,

इस पत्र के पीछे छपी शर्तों पर हम नीचे लिखे विषय, दिनांक और समय पर आपको वार्ता/लघु कथा सहर्ष प्रसारित करेंगे। कृपया संलग्न पुष्टिपत्र पर हस्ताक्षर करके हमको लौटती डाक से भेज दीजिए। इसके लिए हम आपके आभारी होंगे।

We shall be pleased to broadcast your talk(s)/short story on the subject, date and time detailed below on the conditions printed overleaf. We shall be obliged if you could kindly sing and return the attached confirmation sheet not later than ...

शीर्षक ... Title: कॉस्मिक किरणों का मानव शरीर पर प्रभाव।

तारीख (तारीखें) ... Date(s): प्र० तिथि 1-12-91 रिका० 29-11-91 / 2 बजे

प्रसारण का समय ... Time of Broadcast: प्रातः 6:45 बजे

प्रसारण की अवधि ... Duration: 9 मिनट

प्रसारण का स्थान ... Place of Broadcast: आकाशवाणी दरभंगा

फीस ... Fee: 100/- (एक सौ रुपये मात्र)

अनुवर्ती प्रसारण शुल्क ... Subsequent Broadcast fee ...

(इस पत्र के पीछे छपी शर्तों के खण्ड 4(क)/4(ख) एवं 4(ग) के अध्याधीन)

(Subject to Clause 4(a)/4(b) & 4(c) of conditions printed overleaf).

हमारा विशेष निवेदन यह है कि आप कृपया इस शर्त का पालन करके हमारी सहायता करें कि प्रसारण के लिए जो तिथि नियत की गई है उससे कम से कम दस दिन पहले वार्ता/लघु कथा की पांडुलिपि केन्द्र निदेशक के पास पहुँच जाए। शर्त का पालन न होने पर केन्द्र के सामान्य कार्यक्रम में गम्भीर अवरोध पैदा हो जाता है।

We would particularly ask you to assist us by complying with the condition that the manuscript of the talk/short story should be in the hands of the Station Director not less than ten days before the date fixed for the broadcast. The normal routine of the Station is seriously hampered if the condition is not observed.

स्टाम्प शुल्क सरकार द्वारा वहन किया जाएगा।

The Stamp Duty will be borne by the Government.

भवदीय

Yours Faithfully,

25/11

कृते केन्द्र निदेशक

for Station Director

भारत के राष्ट्रपति के लिए और उनकी ओर से

for and on behalf of the President of India.

वाजपेयी

कथा के मामले में लेखक उसका शीर्षक यहाँ भरें।

In case of Short Story the title of the Short Story will be filled in by the author

तार : " आकाशवाणी"
Telegram : "AKASHVANI"

भारत सरकार
आकाशवाणी
Governemnt of India
ALL INDIA RADIO
वार्ता/लघु कथा शाखा
(Talks/Short Story Branch)

आकाशवाणी-कार्यक्रम १
(पैरा ५,१०,११/५, १०, १५ देखिये)
AIR-P-1
(See Paragraph 5, 10, 11/5,10,15)

सेवा में,
To, डा० पी० एन० झा,
............वनस्पतिविज्ञान विभाग, सी० एम० साईंस कॉलेज, दरभंगा
....

सं० विज्ञान वार्ता
No.....................................
दिनांक 18.11.95
Dated.....................19

महोदय/महोदया
Dear Sir/Madam,

इस पत्र के पीछे छपी शर्तों पर हम नीचे लिखे विषय, दिनांक और समय पर आपकी वार्ता/लघु कथा सहर्ष प्रसारित करेंगें । कृपया संलग्न पुष्टिपत्र पर हस्ताक्षर करके हमको लौटाती डाक से भेज दीजिए । इसके लिए हम आपके आभारी होंगें ।

We shall be pleased to broadcast your talk(s)/short story on the subject, date and time detailed below upon the conditions printed overleaf. We shall be obliged if you could kindly sign and return the attached confirmation sheet not later than :-

शीर्षक : वार्ता/वनस्पतिजगत और मौसम
Title..

तारीख (तारीखें) : प्र०- 15.12.95, रिका०- 14.12.95, 2 बजे
Date(s)..

प्रसारण का समय : 7.45 रात्रि
Time of Broadcast..

प्रसारण की अवधि : 10 मि०
Duration of Broadcast..

प्रसारण का स्थान : आकाशवाणी, दरभंगा
Place of Broadcast..

शुल्क रू० : 250/= (दो सौ पचास मात्र)
Fee Rs...

अनुवर्ती प्रसारण का शुल्क :
Subsequent Broadcast..

(इस पृष्ठ के पीछे छपी शर्तों के खण्ड 4 (क)/4(ख) एवं 4 (ग) के अधयधीन)
(Subject to clause 4 (a)/4 (b) & 4 (c) of conditions printed over leaf)

हमारा विशेष निवेदन यह है कि आप कृपया इस शर्त का पालन करके हमारी सहायता करे कि प्रसारण के लिए जो तारीख नियत की गई है उससे कम से कम दस दिन पहले वार्ता/लघु कथा की पांडुलिपी केन्द्र निदेशक के पास पहुँच जाय । इस शर्त का पालन न होने पर कंन्द्र के सामान्य कार्यक्रम में गंभीर अवरोध पैदा हो जाता है ।

We would particularly ask you to Assists us by complying with the condition that the manuscript of the talk/short story should be in the hands of the Station Director not less than 10 days before the date fixed for the broadcast. The normal routine of the Station is seriously hampered if this condition is not observed.

स्टैम्प शुल्क सरकार द्वारा वहन किया जायेगा ।
The Stamp duty will be borne by the Government.

भवदीय,
Yours Faithfully

कृते केन्द्र निदेशक
भारत के राष्ट्रपति के लिए और उनकी ओर से
For and on behalf of the President of India

लघु कथा के मामले में लेखक उसका शीर्षक यहाँ भरे ।
In case of short story the title of the short story will be filled in the author.

LALIT NARAYAN MITHILA UNIVERSITY

KAMESHWARANAGAR, DARBHANGA.

Letter No. 5795-801 Dated-

To

Dr. Mahendra Prasad,
1. ~~Dr. C. C. Saha~~,
Dean, Medical Faculty,
L.N.M.U., Darbhanga.

2. Sri Ram Padarath Singh,
Principal, & Dean, Science Faculty,
B.M.College, Rohika(Madhubani),

3. Dr. C. N. Jha,
Deptt. of Botany,
C.M.Sc.College, Darbhanga;

4. Head,
Univ. Deptt. of Botany/Zoology/Chemistry,
L.N.M.U., Darbhanga;

5. Dr. P. N. Mishra, Convenor
Development Officer,
L.N.M.U., Darbhanga.

Sub:- Constitution of a committee to examine the proposal of the Director, Astrogenetic Society, Darbhanga.

Sir,

The Director, Astrogenetic Society vide his letter dated the 3rd December, 1993 has submitted a proposal alongwith detailed infra-structural requirements for obtaining financial resources to pursue a research oriented voluntary organisation i.e. Astrogenetic Society, Darbhanga. The Vice-Chancellor has been pleased to constitute a committee to examine the proposal and to nominate you a member thereof.

Dr. P.N.Mishra, Development Officer, L.N.M.U., Darbhanga has been nominated as member cum Convenor of the committee concerned.

The Director, Astrogenetic Society, Darbhanga will present statistics of the works done so far, future programmes etc. alongwith the bye-laws of the society, list of office bearers etc.

Kindly intimate your consent to the Convenor of the committee at an early date.

Yours faithfully,

(J. R. Singh) 13/5/94
REGISTRAR

LALIT NARAYAN MITHILA UNIVERSITY

KAMESHWARANAGAR, DARBHANGA.

Letter No. 24602 / Dated. 01-11-93 /

To

Dr. C. N. Jha,
Department of Botany,
C.M.Sc.College, Darbhanga
-cum-
Director, Astrogenetic Society,
Kathalbari, Darbhanga.

Sub:- Proposed research with Astrogenetic approach.

Sir,

Your letter dated 16.9.93, on the subject ~~refer~~ referred to above. In this connection, I am directed to request you kindly to let us have a list of infra-structural requirements for the needful.

Yours faithfully

(R.K.Jha)

DEPUTY REGISTRAR(I)

Hari/-
1_11_93

(२) हिन्दुस्तान, पटना, शनिवार, १५ अप्रैल १९९५ ई.

एस्ट्रोजेनेटिक्स नामक नयी चिकित्सा पद्धति का इजाद

(एक प्रतिनिधि)

दरभंगा, १४ अप्रैल। शोध और चिंतन के आधार पर रोग के निदान हेतु विश्व में कई पद्धतियों को प्रचलित किया गया है और इससे मानव कल्याण का मार्ग प्रशस्त हुआ है। इसी श्रृंखला में विज्ञान विषय के एक स्थानीय शिक्षक ने अध्ययन, अनुभव व अनुसंधान के आधार पर 'एस्ट्रोजेनेटिक्स' नाम से एक नयी चिकित्सा पद्धति का इजाद किया है जिससे असाध्य रोगों का भी इलाज करने का दावा किया गया है। अब तक विभिन्न रोगों से ग्रस्त सैकड़ों मरीजों का इलाज इस पद्धति से किया गया है तथा उपलब्धि ६० से ९० प्रतिशत तक आंकी गयी है।

इस पद्धति को विकसित व प्रचारित करने के लिए गठित 'एस्ट्रोजेनेटिक्स सोसाइटी' द्वारा देश व विदेश के वैज्ञानिक संस्थाओं से पत्राचार व सूचनाओं के आदान-प्रदान करने के बाद अखिल भारतीय आयुर्विज्ञान संस्थान, इंडियन कौंसिल आफ मेडिकल रिसर्च तथा साइंस कौंसिल आफ जापान ने इस विधि को विकसित करने में रुचि दिखाई है। सोसाइटी से प्राप्त पत्र व सूचनाओं को साइंस कौंसिल आफ जापान के अध्यक्ष डा. जीरो कोंडो ने रिसर्च लियाजों कमेटी आन बॉटनी के अध्यक्ष प्रो. इवशे तकहाशि को अध्ययन कर मंतव्य देने के लिए कहा है। इसी तरह आयुर्विज्ञान संस्थान में पेडियेट्रिक्स के प्रोफेसर डा. आई.सी. वर्मा ने वैसे तीस मरीजों के विषय में जानकारियां मांगी, है जो संस्थान के द्वारा रोगमुक्त नहीं हो सके लेकिन उन मरीजों का इलाज इस नयी विधि से किया जा सका है। वांछित सूचनाएं भेजी गयी हैं।

इस बीच ललित नारायण मिथिला विश्वविद्यालय के पूर्व आयुक्त कुलपति आलोक चंद्र रंजन ने विकास पदाधिकारी डा. पशुपति नाथ मिश्र के संयोजकत्व में एक उच्चस्तरीय समिति का गठन किया तथा उससे कहा गया कि सोसाइटी द्वारा प्रदत्त सूचनाओं का अध्ययन कर शोधकार्य के लिए आधारभूत सुविधा व आर्थिक सहायता के लिए प्रस्ताव तैयार करें। समिति के अन्य सदस्यों में चिकित्सा संकायाध्यक्ष, विज्ञान संकायाध्यक्ष के साथ-साथ स्नातकोत्तर प्राणि शास्त्र, जीव विज्ञान व रसायन शास्त्र के विभागाध्यक्षों के अतिरिक्त स्थानीय चंद्रधारी मिथिला विज्ञान महाविद्यालय के प्राणिशास्त्र विभाग के उपाचार्य डा. जगन्नाथ झा शामिल हैं।

सोसाइटी के कागजात से ज्ञात होता है कि प्रदेश के कोने-कोने से तो मरीज आते ही हैं लेकिन इसके अलावा उत्तर प्रदेश के कई क्षेत्रों से भी मरीज यहां आकर अपने असाध्य रोगों का इलाज करवाते हैं। कैंसर, दमा, एलर्जी, गैस्ट्रिक, मधुमेह, चर्मरोग, जौण्डिस, स्त्रीजनित रोग, गठिया, वात, स्पौंडलाइटिस आदि रोगों का भी इलाज इस पद्धति से किया जा चुका है।

सम्पर्क करने पर सोसाइटी के संस्थापक तथा इस पद्धति के विशेषज्ञ डा. सी.एन. झा, जो स्थानीय सी.एम. साइंस कालेज में प्राणिशास्त्र के उपाचार्य हैं, ने कहा कि सर्वप्रथम १९८१ में शोधकार्य के समय अचानक उनके दिमाग में यह प्रश्न उठने लगा कि क्या सूर्य व सितारों के कारण इन सशक्त ऊर्जापुंज से घिरे प्रत्येक मानव इनसे अछूते हैं? इसी से प्रेरणा पाकर उन्होंने इस विषय पर शोधकार्य प्रारम्भ किया और अंततः 'एस्ट्रोजेनेटिक पद्धति' का विकास हुआ। उन्होंने कहा कि प्रथम चरण के परिणाम काफी उत्साहवर्धक रहे और इसके आधार पर यह सिद्ध हुआ कि रोगों का मूल प्राथमिक कारण कास्मिक किरण ही है। इस तरह रत्नों, उपरत्नों की सहायता से शरीर पर पड़ रहे कास्मिक किरणों को संतुलित कर विभिन्न रोगों का इलाज किया जाने लगा और आशातीत सफलता मिली।

डा. झा ने बताया कि इस पद्धति की लोकप्रियता को दृष्टि में रखते हुए १९८५ में निःशुल्क मानव सेवा संस्थान के रूप में 'एस्ट्रोजेनिक सोसाइटी' का गठन हुआ और कई युवक-युवतियां आकर इस मानव कल्याणी शोध प्रणाली से रोग उन्मूलन यज्ञ में आहुति प्रदान करने लगे। उनके मुताबिक करीब २५ प्रतिशत रोग आयुर्विज्ञान द्वारा अपरिचित रह जाते हैं और उनके सूक्ष्म उपकरणों में भी नहीं उतरते। करीब २५ प्रतिशत रोग उनके कार्यक्षेत्र से बाहर हैं जिनके कारणों तक को वे समझ नहीं पाते हैं। इसका अर्थ हुआ कि मात्र ५० प्रतिशत रोगों का निदान आयुर्विज्ञान पद्धति में है लेकिन इस नयी पद्धति में सारे रोगों का उपचार होता रहा है। इसके अलावा यह प्रक्रिया विशेष रक्षाकवच की भी भूमिका अदा करती है।

आम मरीजों के बीच 'स्टोन थिरैपी पद्धति' से मशहूर यह प्रक्रिया कई युवकों व युवतियों को इसके अध्ययन के लिए आकृष्ट करने लगी और तब १९८५ से इस संस्थान द्वारा पठन-पाठन का कार्यक्रम भी प्रारम्भ किया गया और एक वर्ष में सैद्धांतिक व प्रायोगिक अध्ययन समाप्त कर कई लोगों ने इस पद्धति से इलाज करना प्रारम्भ किया है। अभी तक पटना, बोकारो, समस्तीपुर आदि कुछ जगहों पर निःशुल्क चिकित्सा उपकेंद्र खोले गये हैं और सैकड़ों मरीजों का इलाज किया जा चुका है।

एक प्रश्न के जवाब में डा. झा ने कहा कि धन व शोधागार सुविधा के अभाव में शोधकार्य में अपेक्षित गति नहीं आ रही है और इस कारण देश व विदेश की वैज्ञानिक संस्थाओं से सम्पर्क करने में भी कठिनाइयों का सामना करना पड़ता है। सोसाइटी इस समस्या को दूर करने में लगी है।

जब उनसे यह पूछा गया कि क्या यह पद्धति जेनेटिक इंजीनियरिंग में सहयोगी सिद्ध हो सकती है तो जवाब हुआ कि वस्तुतः जेनेटिक इंजीनियरिंग जीन प्रत्यारोपण पद्धति पर आधारित है, क्योंकि आधुनिक वैज्ञानिक तकनीक द्वारा जीन के कार्यक्षेत्र में हस्तक्षेप सम्भव नहीं है लेकिन एस्ट्रोजेनेटिक तकनीकों द्वारा जीन के कार्यक्षेत्र में हस्तक्षेप सम्भव होता है जिससे बिना प्रत्यारोपण के खास हद तक जेनेटिक इंजीनियरिंग का उद्देश्य पूरा हो जाता है। इस प्रक्रिया में वस्तुतः रत्न प्रभावी होते हैं। उदाहरण के रूप में उन्होंने बताया कि इंसुलीन की कमी वाले रोगी को इनसुलीन निर्माण चालू करना, थायराइड की गड़बड़ी वाले रोगी को उससे मुक्ति दिलाना, इसी प्रकार मेंटल रिटार्डेशन एवं कई अन्य आनुवांशिक गुणों को विकसित कर यह आधुनिक जेनेटिक इंजीनियरिंग की दृष्टि से काफी अधिक उपयोगी सिद्ध हुआ है।

स्थानीय बुद्धिजीवियों ने इस पद्धति पर प्रतिक्रिया व्यक्त करते हुए कहा कि अभी इसमें बहुत अधिक शोध की आवश्यकता है और इसलिए देश के प्रमुख वैज्ञानिकों को इस प्रतिपादित सिद्धांत को और विकसित करने में अपना योगदान देना आवश्यक है ताकि मानव कल्याण में यह भी एक मील के पत्थर की तरह सहायक सिद्ध हो।

"दैनिक जागरण" 18 Oct 2000

रत्नों से लाइलाज रोगों का उपचार सम्भव

दरभंगा, 18 अक्टूबर (स.स.)। आकाशीय किरणों का प्रभाव मानवीय शरीर पर पड़ता है, जिससे शरीर का आंतरिक वातावरण नियंत्रित होता है। हजारों रोगियों पर किये गये अब तक के शोधों से यह जाहिर हो चुका है कि उनके आंतरिक गुणों पर आकाशीय किरणों का प्रभाव होता है। जिसका निराकरण रत्नों एवं उपरत्नों से हो सकता है। यह बातें प्रसिद्ध रत्न चिकित्सक डॉ. सी.एन. झा ने कही। वे मब्बी में आयोजित भव्य रत्न प्रदर्शनी एवं अन्तर्राष्ट्रीय ज्योतिष शोध और रत्न चिकित्सा केन्द्र के उद्घाटन के मौके पर बोल रहे थे। इसका आयोजन बागमती जेम्स मेन्युफेक्चरिंग कम्पनी ने किया।

उन्होंने बताया कि वर्ष 1983 में एस्ट्रोजेनेटिक्स सोसाइटी की स्थापना उन्होंने की थी, जिसमें अब तक लगभग 30 हजार मरीजों का इलाज किया गया है। इसमें 10 हजार ऐसे मरीज थे जिसे बड़े-बड़े चिकित्सकों ने नकार दिया था। उन्होंने कहा कि पेप्टाइटिस 'बी' और एड्स एवं अन्य जानलेवा व लाइलाज बीमारियों से बचाव में भी रत्न सहायक हो सकता है। डॉ. झा ने दावा किया कि जापान के साइंस काउन्सिल और दिल्ली में स्थित आल इंडिया मेडीकल इंस्टीच्यूट आफ मेडीकल साइंस ने भी उनके शोध में रुचि दिखाते हुए उसे मान्यता दिया है। इस दिशा में बागमती जेम्स मेन्यूफैक्चरिंग कम्पनी के योगदान की भी उन्होंने सराहना की। समारोह का उद्घाटन ज्योतिष विभागाध्यक्ष पं. रामचन्द्र झा ने किया। जबकि कई प्रमुख ज्योतिषों संतोष जी, राजीव ज्योतिषी, कलीकान्त झा, विष्णुधर झा सहित मिथिला एवं दूरदराज से आये नामीगिरामी चिकित्सकों, अध्यापकों आदि ने भी अपना-अपना विचार व्यक्त किया। धन्यवाद ज्ञापन धीरेन्द्र मिश्र ने किया।

आरोप : ल.ना. मिथिला विश्वविद्यालय प्रशासन पर एम.बी.ए. एवं डेंटल कालेज की परीक्षा आयोजित करने तथा सामान्य शिक्षा से जुड़े परीक्षाओं की वैकल्पिक व्यवस्था नहीं करने के लिए दोरंगी नीति अपनाने का आरोप लगाया गया है। जिला युवा जनता दल (यू) के उपाध्यक्ष सुशील कुमार चौधरी ने जारी बयान में कहा है कि निजी महाविद्यालयों में पढ़ने वाले धनाढ्य छात्रों को खुश करने तथा सामान्य कालेजों में अध्ययनरत छात्रों की उपेक्षा विश्वविद्यालय द्वारा किया जा रहा है।

स्पष्टीकरण : सी.एम. विज्ञान कालेज में दरभंगा राज कोटे पर नामांकन को लेकर प्रधानाचार्य से स्पष्टीकरण पूछा गया है। अमित कुमार अमन ने कुलपति को एक ज्ञापन देकर राज कोटे में अपना नामांकन का दावा किया है।

रत्न पत्थरोंका मानव चिकित्सा पर सार्थक परिणाम : डा. झा

(दरभंगा कार्यालय)

दरभंगा, 18 अक्तूबर। आकाशीय किरणोंका प्रभाव मानव शरीर पर पड़ता है जिससे शरीरका आंतरिक वातावरण नियंत्रित होता है। हजारों रोगियों पर किये गये अब तकके शोधोंसे जाहिर हो चुका है कि रोगियोंके आंतरिक गुणोंका कारण आकाशीय किरणें हैं जिसका निराकरण रत्नों एवं उपरत्नोंके द्वारा हो सकता है।

उपर्युक्त बातें रत्न चिकित्सक डा. सीएन झाने बागमती जेम्स शोध एवंरत्न चिकित्सा केन्द्रके उद्‌घाटनके अवसर पर कही। उन्होंने बताया कि 1983 में एस्ट्रोजेनटिक्स सोसाइटी का स्थापना कर करीब तीस हजार रोगियोंका इलाज किया है। इसमें करीब दस हजार ऐसे रोगी थे जिन्हें मेडिकल संस्थानोंने नकार दिया था।

श्री झाने कहा कि हेपेटाइटिस या एड्स जैसी जानलेवा बीमारीका इलाज करनेके साथ ही मनोवैज्ञानिक, मेडिकल, जेनेटिक्स या क्रानिक डीजिज, जो लाइलाज है हम उसे भी ठीक कर सकते हैं। रत्न पत्थरोंके चिकित्सकीय उपयोगकी दिशामें जापानके साइंस काउंसिल एवं दिल्लीके आल इंडिया मेडिकल इन्सटीच्यूट ऑफ मेडिकल साइंसने भी रूचि दिखायी है।

दरभंगामें अपने तरहका यह पहला रत्नोंका प्रदर्शनी था। इसका उद्‌घाटन ज्योतिषीय विभागाध्यक्ष पंडित रामचन्द्र झा ने किया। दरभंगा–मुजफ्फरपुर मार्गके मब्बीमें स्थापित यह संस्थान दरभंगाके लिए अनोखा है। इस मौके पर ज्योतिष संतोष जी, राज़ीव ज्योर्तियोगी, कालिका दत्त झा, विष्णुधर झा आदि मौजूद थे। प्रदर्शनीमें असली एवं नकली रत्नोंके पहचान की विधि भी बताया गया।

'आज' 19 Oct 2

(A) Format of Prescription used for Patients

(B) Few Samples of Patients Comment.

(C) Few Samples of under taking forms (u) used.

(D) Few Volumes of Patients Diary (36) Still safe.

(A)

"HELP ME TO HELP YOU"

ASTROGENETIC SOCIETY

H. O. :- KATHALBARI, DARBHANGA

(Out Door/In Door)

Ref. No. Date

Name :—

Weight :—

Disease :—

1. Moonga + Garnet + Topaz + Moti

(3, 5, 7 days) R. A.

2. Lajbrat + Gomed + Tigers eye

L. A -

3. Onex + Asfatic

Neck—

Precautions –

1. Report after each 10/15 days.
2. No antibiotic drug after 4 dozes.
3. Dozes should not be let loose.

Pokhraj A. D. Others

(B)

ASTRO-GENETIC SOCIETY

KATHALWARI

DARBHANGA - 846004

DISEASE ERADICATION PROGRAMME:

I, ...K P Sinha.... have been treated by "ASTRO-GENETIC SOCIETY" through "STONE - THERAPY" and felt ..70-80..%/complete recovery in ...Allergic manifestations including bronchial asthma due to cold, dust, pollen etc. ... disease/diseases within ..2-3.... days/months.

K P Sinha
Signature

Date: 31.1.91.
Reference No. 5/83-06
Place: Darbhanga
Address: Digha (west) Professors' Colony Darbhanga - 846004

K.P. Sinha
Reader
Deptt. of Zoology
C.M. Sc. College
Darbhanga

I do hereby agree with the principles and objects of ASTRO-GENETIC SOCIETY. It is a boon for mankind perfectly under scientific discipline. Its inclusion in Life Science should be enevitable. Free service rendered by this Society establishes human values too.

K P Sinha
Signature

Date: 31.1.91

ASTRO-GENETIC SOCIETY

KATHALWARI

DARBHANGA - 846004

DISEASE ERADICATION PROGRAMME:

I, ...Puspa Devi...................................... have been treated by "ASTRO-GENETIC SOCIETY" through "STONE - THERAPY" and felt. ....%/complete recovery in ......Heart trouble, Gastric..........
....
disease/diseases within ....20....... days/~~month~~s.

पुष्पा देवी

Date: 17.7.91 Signature

Reference No.

Place:

Address: W/o Upendra Kr. Lal Das
Assistant
~~Exam~~ Dy. Registrar
L.N.M.V.
Darbhanga

I do hereby agree with the principles and objects of ASTRO-GENETIC SOCIETY. It is a boon for mankind perfectly under scientific discipline. Its inclusion in Life Science should be enevitable. Free service rendered by this Society establishes human values too.

पुष्पा देवी

Date: Signature

ASTRO-GENETIC SOCIETY

KATHALWARI

DARBHANGA - 846004

DISEASE ERADICATION PROGRAMME:

I, ...Amita Gupta.................................. have been treated by "ASTRO-GENETIC SOCIETY" through "STONE - THERAPY" and felt ....%/complete recovery in ...Leucorrhia, fainting tendency, headache, sex perversion etc. .... disease/diseases within .....15........ days/months.

Date: 3.2.91

Signature

Reference No.

Place: DARBHANGA

Address: c/o Nanu Ram
Baijnath
Bara Bazar
Katihar

I do hereby agree with the principles and objects of ASTRO-GENETIC SOCIETY. It is a boon for mankind perfectly under scientific discipline. Its inclusion in Life Science should be enevitable. Free service rendered by this Society establishes human values too.

Date: 3.2.91

Signature

Form-U 10

एस्ट्रोजेनेटिक सोसाइटी, कटहलबाड़ी, दरभंगा Ref. No. 78/19

चिकित्सा आवेदन-पत्र

मैं पाल्य श्री/श्रीमती ध्रुव प्रसाद पता [illegible] दरभंगा विगत .... दिनों से .... व्याधि ग्रसित हूँ एवं अन्य सभी पद्धति से इलाज करवा चुका/चुकी हूँ। किसी प्रकार का लाभ नहीं होने के कारण मैं/पाल्य इस सोसाइटी द्वारा अपनायी गई पद्धति से चिकित्सा करवाना चाहता/चाहती हूँ जिसका सारा उत्तरदायित्व स्वयं मुझ पर होगा।

रोग/रोगों का ब्योरा

नाम	उम्र	रोग का नाम	पूर्व चिकित्सा का विवरण	विशेष
ध्रुव प्रसाद	32	सर्दी, खाँसी, दमा	गणपति मिश्रा से एलोपैथिक एवं होमियोपैथ आयुर्वेद और [illegible] होमियोपैथ वाले [illegible] से उपचार करवाये	

[illegible]

आवेदक का हस्ताक्षर

(C) Form-U

Ref. No. 3/21 25.1.93

स्ट्रैटेजिक सोसाइटी, कटहलवाड़ी, दरभंगा

चिकित्सा आवेदन-पत्र

मैं पाल्य श्री/श्रीमती पार्वती देवी पता ग्रा० + पो० हरपुर एलौथ

जिला - समस्तीपुर विगत 25 दिनों से किडनी रोग से ग्रसित हूँ एवं अन्य सभी पद्धति से इलाज करवा चुका/चुकी हूँ। किसी प्रकार का लाभ नहीं होने के कारण मैं/पाल्य इस सोसाइटी द्वारा अपनायी गई पद्धति से चिकित्सा करवाना चाहता/चाहती हूँ जिसका सारा उत्तरदायित्व स्वयं मुझ पर होगा।

रोग/रोगी का ब्यौरा

नाम	उम्र	रोग का नाम	पूर्व चिकित्सा का विवरण	विशेष
पार्वती देवी	47 वर्ष	किडनी का असफल होना	एलोपैथी में ... डी०एम० ... आर०ती० ... तथा ... होमियोपैथी चिकित्सा ... पटना ... किया गया।	स्थिति ... के बारे

सुशील प्रसाद राय

आवेदक का हस्ताक्षर

24.1.93

Form-U

2
SR

सूदी मैग्नेटिक पोलारटी ..., दार्जिलिंग

चिकित्सा-आवेदन-पत्र

मैं प्रार्थी श्री/श्रीमती .... पता .... विगत .... दिनों से .... रोग से ग्रसित हूँ एवं अन्य सभी पद्धतियों का इलाज करवा चुका/चुकी हूँ। किसी प्रकार का लाभ नहीं होने के कारण मैं/प्रार्थी इस पोलारटी द्वारा अपनायी गई पद्धति से चिकित्सा करवाना चाहता/चाहती हूँ जिसका सारा उत्तरदायित्व स्वयं मुझ पर होगा।

रोग/रोगी का ब्योरा

नाम	उम्र	अवधि	पूर्वचिकित्सा का विवरण	विशेष
[illegible]	34 वर्ष			

[illegible]

आवेदक का हस्ताक्षर

[illegible]

Form-U

9
SR—35/19

स्ट्रोबैजेटिक सोसाइटी, कटहलवाड़ी, दरभंगा

चिकित्सा-आवेदन-पत्र

मैं पात्र श्री/श्रीमती अरुण कुमार झा .. पता ग्राम .. [illegible] पो० [illegible] ([illegible]) विगत बचपन से ... दिनों से रोग से ग्रसित हूँ एवं अन्य सभी पद्धतियों से इलाज करवा चुका/चुकी हूँ। किसी प्रकार का लाभ नहीं होने के कारण मैं/पात्र इस सोसाइटी द्वारा अपनायी गई पद्धति से चिकित्सा करवाना चाहता/चाहती हूँ जिसका सारा उत्तरदायित्व स्वयं मुझ पर होगा।

रोग/ रोगी का ब्यौरा

नाम	उम्र	रोग का नाम	अवधि	पूर्वचिकित्सा का विवरण	विशेष
अरुण कुमार झा	30 वर्ष	हृदय रोग	30 वर्ष से	[illegible] जी.बी.पी.एच New Delhi	प्रोलैप्स [illegible]

अरुण कुमार झा
आवेदक का हस्ताक्षर

Form-U

5

Ref.No.—

एस्ट्रोजेनेटिक्स सोसायटी, कटहलवाड़ी, दरभंगा

चिकित्सा-आवेदन-पत्र

मैं प्रार्थी श्री/श्रीमती ..आशा देवी.................. पता ..सीतामढ़ी.... ..दरभंगा.... विगत ..कमर दर्द... दिनों से रोग से ग्रसित हूँ एवं अन्य सभी पद्धतियों से इलाज करवा चुका/चुकी हूँ। किसी प्रकार का लाभ नहीं होने के कारण मैं/प्रार्थी इस सोसायटी द्वारा अपनायी गई पद्धति से चिकित्सा करवाना चाहता/चाहती हूँ जिसका सारा उत्तरदायित्व स्वयं मुझ पर होगा।

रोग/रोगी का ब्यौरा

नाम	उम्र	रोग का नाम	अवधि	पूर्व चिकित्सा का विवरण	विशेष
आशा देवी	30 वर्ष	कमर दर्द		सीतामढ़ी	

आशा देवी

आवेदक का हस्ताक्षर

Form-U

9
Ref no.

स्ट्रो जैनेटिक सोसाइटी, कटहलवाड़ी, दरभंगा

चिकित्सा-आवेदन-पत्र

मैं पाल्य श्री/श्रीमती रेणुका देवी .... पता ग्रा+पो० हरपुर एलौथ
लगुनिया जिला- समस्तीपुर विगत ५ वर्षों .... दिनों से ....
रोग से ग्रसित हूँ एवं अन्य सभी पध्दतियों से इलाज करवा चुका/चुकी हूँ। किसी प्रकार का लाभ नहीं होने के कारण मैं/पाल्य इस सोसाइटी द्वारा अपनायी गई पध्दति से चिकित्सा करवाना चाहता/चाहती हूँ जिसका सारा उत्तरदायित्व स्वयं मुझ पर होगा।

रोग/रोगी का ब्योरा

नाम	उम्र	अवधि	पूर्वचिकित्सा का विवरण	विशेष
रेणुका देवी	२८	५ वर्षों	समस्तीपुर मुजफ्फरपुर पटना	

आवेदक का हस्ताक्षर

Rai
1/9/92

ऐस्ट्रोजेनेटिक सोसायटी कटहलबाड़ी, दरभंगा के तत्वावधान में 10 वें वर्ष मानवता (Humanity) दिवस के पावन अवसर पर संस्था (Voluntary organisation) की ओर से "Service to man is service to God" की परिकल्पना को प्रचारित-प्रसारित करने हेतु संस्था के कुछ मूल वैज्ञानिक तथ्य संस्थापक की ओर से जन जागरण के लिए प्रस्तुत किया गया।

यह अतिशयोक्ति नहीं कि इस धरातल पर आये प्रत्येक मानव को इस दिव्य परन्तु वैज्ञानिक तथ्य की जानकारी होनी चाहिए।

आएं हम मानवतावादी धर्म के पुर्नस्थापना के लिए दृढ़ संकल्पित हों।

ऐस्ट्रोजेनेटिक्स का इतिहास

सन 1983 ई. के श्रावण शुक्ल द्वादशी के सायंकाल मैं अपने छोटे मुक्ताकाश उद्यान में बैठकर कभी आँखें मुंदकर गैलेक्सीज की कल्पना करता रहा फिर आँखें खोलते हुए चाँद-सितारों के जगमगाता पटल मानो अपने ऊर्जा-पूंज से आकस्मिक दिव्य ज्ञान की ओर खीचने लगा। फिर ऐस्ट्रोफिजिक्स की वह सत्य खोज, क्या इन सशक्त ऊर्जा-पूंज से घिरे प्रत्येक मानव इनसे अछूते हैं? क्या जीवन विज्ञान इस पर सोचता है? नहीं तो फिर इस अनदेखी का कारण? जीवन विज्ञान का संक्षिप्त इतिहास या उसका अनुसंधान प्रणाली? ये ऊर्जा-पूंज कथमपि निरर्थक नहीं हो सकते। फिर इलेक्ट्रोमैग्नेटिव वेब के चमत्कार से जीवन विज्ञान तो प्रभावित ही है। इन रहस्यों के उद्‌घाटन और उनके उधेड़बुन में पड़ा मैं घंटो बाद कल्पना जगत में खोए-खोए मानसिक पटल पर एक शब्द को लेखांकित किया 'ऐस्ट्रोजेनेटिक्स''। उसी दिन से अपने जीव विज्ञान के सहयोगी डॉ0 रघुनाथ प्रसाद सिंह के साथ इसकी चर्चा, फिर इसके वैज्ञानिक सार्थकता की खोज पर काफी सक्रियता से हमलोग जुट गए। सिद्धांतो की श्रृंखला बंधी फिर शोध के कार्यक्रम एवम् प्रक्रिया पर विचार होने लगे।

चूँकि इस गैलेक्सी में तिल भर भी इन किरणों से मुक्त स्थान सृजित करना संभव नहीं है अतएव किरणों को बढ़ाकर ही शोध प्रक्रिया स्थापित करने का विचार किया गया। शोध का निर्देशन ''प्रभाव से कारण'' प्रणाली पर स्थापित की गयी क्योंकि कारण करोड़ों मील दूर से आते किरण-पूंज है। शोध का प्रयोज्य अतिविकसित प्राणी मानव शरीर को लिया गया। किरणों की सांद्रता बढ़ाने-घटाने हेतु शुद्ध किरणों द्वारा निर्मित किरणग्राही पदार्थों (रत्न, उपरत्न) को लिया गया। प्रथम चरण के परिणाम काफी उत्साहवर्द्धक रहे। इसके आधार पर यह सिद्ध हुआ कि रोगों का मूल प्राथमिक कारण कॉस्मिक किरण ही है। इसकी सूचना विश्व के मुख्य वैज्ञानिक संस्थाओं को दी गयी। साईंस काउंसिल जापान ने इस विषय पर अपने एक विशेष समिति का गठन किया एवम् अपने निर्णय द्वारा हमें इस शोध के लिए प्रोत्साहित किया। इस विषय की सूचना ऑल इंडिया इंस्टीच्यूट ऑफ मेडिकल साईंस, यू0 जी0 सी0 आई0 सी0 एम0 आर0 डी0 एस0 टी0 आदि भारतीय

संस्थानों को दी गई। इसमें ऑल इंडिया इंस्टीच्यूट ऑफ मेडिकल साईंस और आई0 सी0 एम0 आर0 ने विशेष रूचि दिखायी।

1985 ई0 में निःशुल्क मानव सेवा संस्थान के रूप में ऐस्ट्रोजेनेटिक्स सोसायटी का गठन हुआ और कई युवक-युवतियाँ आकर इस मानव कल्याण शोध प्रणाली से रोग-उन्मूलन यज्ञ में आहूति प्रदान करने लगे। असंख्य शोधों का सिलसिला चलता रहा। हजारों कराहते, 'रोग-शैय्या' पर पड़े निःसहाय मानवता को राहत दी गई। ऐसा पाया गया कि करीब 25 फीसदी रोग आयुर्विज्ञान द्वारा अपरिचित रह जाते हैं। उनके सूक्ष्म उपकरणों में भी नहीं उतरते; 25 फीसदी रोग उनके कार्य-क्षेत्र से बाहर हैं जिसके कारणों तक को वे नहीं समझ पाए हैं अर्थात् मात्र 50 फीसदी रोगों का निदान वहाँ संभव है जबकि हमारी इस प्रक्रिया द्वारा सारे रोगों का निदान चलता रहा है। इतना ही नहीं यह प्रक्रिया विशेष रक्षा-कवच की भी भूमिका अदा करती है।

अनेक सुझाव प्राप्त होने के बाद 1987 ई0 में इस संस्थान द्वारा पठन-पाठन का कार्यक्रम भी निर्धारित हुआ। एक वर्ष की अवधि में सैद्धांतिक एवम् प्रायोगिक अध्ययन समाप्त करनेवाले अध्ययनकर्त्ता को अलग-अलग स्थानों पर केन्द्र खोलने एवं मानवता की निःशुल्क सेवा करने के लिए प्रेरित किया गया। इस क्रम में बोकारो में प्रो0 नृपेन्द्र नाथ झा, लोहना में श्री विनोदा नन्द झा, लहेरियासराय में डॉ0 विनय कुमार सिंह, अनिल कुमार झा, समस्तीपुर में मोतीचन्द आदि के द्वारा ऐस्ट्रोजेनेटिक्स केन्द्रो का संचालन किया जा रहा है। अनेको छात्र-छात्राएँ अध्ययनरत हैं जो मानवता की सेवा के लिए कृतसंकल्पित हैं। छात्राओं में अल्पना अग्रवाल, सरोज सिंह आदि मुख्य हैं।

शोध संस्थान सिर्फ रोग उन्मूलन तक ही सीमित नहीं है बल्कि इसके कार्य क्षेत्र के अन्तर्गत कई सामाजिक विषय-वस्तु भी सन्निहित है। सामाजिक कार्य क्षेत्र के तहत विधवापन-उन्मूलन, अपराधकर्मियों में सुधार, वेश्यावृति में सुधार, नशाखोरों में सुधार आदि प्रमुख रूप से लिए गए हैं।

मानवतावादी धर्म पर आधारित यह संस्था एक ओर विज्ञान की चरमसीमा दूसरी ओर धर्म की यथार्थवादिता दर्शाती है।

आइये, नव वर्ष 1992 के अवसर पर हम सब पीड़ित मानवता की सेवा के लिए एक बार फिर कृत संकल्पित हों।

ज्ञातव्य है कि ऐस्ट्रोजेनेटिक सोसायटी मेरे ही निवास पर संचालित होता था अतः दूर-दराज से आये रोगी घंटो, कभी-कभी दो-तीन दिन यहीं रुक जाते थे। इस प्रकार पीड़ित मानवता की अन्य आवश्यकताओं की पूर्ति घर में उपस्थित सभी लोग कुछ न कुछ किया करते थे, तभी इस पुनीत महायज्ञ का इतने लम्बे समय (25-30 वर्ष) तक संचालन संभव हुआ। सिर्फ रोगी ही नहीं अपितु अहर्निश सेवा में संलग्न संस्था के सदस्यगणों के

आवश्यकताओं की पूर्ति भी काफी ध्यानपूर्वक किया करते थे, जिसमें मुख्य रूप से श्रीमती कुमुद झा, प्रीतिश्री, प्रज्ञाश्री, संगीत कुमार 'आशीष', शिव नाथ झा, शम्भु नाथ झा आदि।

संस्था उन महापुरुषों को कृतज्ञता के साथ श्रद्धासुमन अर्पित करती है; जिन्होंने संस्थापक पुत्र से अपने हितों को त्याग कर मानवता यज्ञ के लिए यज्ञ-परामर्श अपनाया। ऐसे महामानव पिता स्व. कुलानन्द झा एवम् पुनेश्वरी देवी को शत शत नमन।

डॉ0 सी0 एन0 झा
संगीत सदन, कटहलबाड़ी, दरभंगा
बिहार, भारतवर्ष

कॉस्मिक किरणें क्या हैं?

सन 1912 ई0 में विक्टर हेस एवं उनके सहयोगियों ने गुब्बारा उड़ान द्वारा 16 हजार फीट पर सशक्त ऊर्जा-पूंज आयोनाइजेशन मीटर द्वारा प्राप्त किया। वे इस तथ्य पर पहुँचे कि इस सशक्त ऊर्जा का स्रोत सूर्य, तारे एवं नीहारिकाएँ अर्थात् दिन एवं रात का दैदिप्यमान आकाश है जो पृथ्वी को चारों ओर (360°) से घेरे हुए है। ये किरणें सशक्त ऊर्जा कण से निर्मित छोटी तरंग वाली होती है, जिनकी ऊर्जा क्षमता एक से एक लाख वेभस्ट्रन तक पायी गई। करोड़ों मील दूर से चलकर ये किरणें सर्वप्रथम ओजोन सतह पार करती है। इस क्रम में इनकी विशेष रासायनिक संरचना एवं खास शीतोष्ण प्रभाव वाले ग्रहों से प्रतिक्रिया होती है। प्रतिक्रिया एवं अन्र्तप्रतिक्रिया के फलस्वरूप ये अनन्त किस्मों की हो जाती है। पृथ्वी के गुरुत्वाकर्षण के निकट कुछ किरणें विकर्षित हो जाती है तथा कुछ पृथ्वी के वातावरण में प्रविष्ट होती है। इनका प्रत्येक कण पृथ्वी के वातावरण से प्रतिक्रिया कर किरणों की भयावह बृष्टि उत्पन्न करते हुए पृथ्वी के कण-कण में समा जाता है। ऐसा समझा जाता है कि इन किरणों का योगदान रासायनिक एवं जैविक विकास में प्राथमिक रूप से रहा होगा और है। वैज्ञानिक स्तर पर ये प्रोटोन होते हैं जो सभी ग्रहों एवं ब्रह्माण्ड में व्याप्त हैं। यह ब्रह्मण्ड में पायी जाने वाली एक विलक्षण और अकेली शक्ति है। हमारे गैलेक्सी में जो किरणें पृथ्वी के गुरुत्वाकर्षण बल से पृथ्वी तक पहुँचती है, उनके द्वारा पृथ्वी की चारों ओर एक विशेष प्रकार के वातावरण का निर्माण होता है, जिससे जैविक विकास संभव हो पाता है।

जैविक विकास में अन्य ग्रहों की तुलना में पृथ्वी की अपनी विशिष्ट प्रकार की रासायनिक संरचना की भूमिका महत्वपूर्ण है। इस सत्य से इनकार नहीं किया जा सकता है कि पृथ्वी की इस विशिष्ट रासायनिक संरचना पर कॉस्मिक किरणों की प्रतिक्रिया के फलस्वरूप ही यहाँ जैविक विकास हुआ है। अतः जिस प्रकार इन क्रिया-कलापों में कॉस्मिक किरणों का महत्व प्राथमिक है उसी प्रकार सभी तरह की प्राकृतिक प्रक्रियाओं में इनकी भूमिका प्राथमिक एवं वातावरण की भूमिका माध्यमिक है; जिसे आम तौर पर प्राथमिक समझ लिया जाता है।

वस्तुतः तथाकथित सौर-ऊर्जा कॉस्मिक किरणों का मिश्रित रूप है, जिसका मात्र लाल भाग सूर्य से आता है। अन्य सभी रंग के किरणों का उद्‌गम-स्थान आकाशखण्ड है; जिनके कारक तारें और नीहारिकाएँ हैं। किरणों में संचालन-क्षमता के अलावा वातावरण-सृजन, उत्परिवर्त्तन आदि प्रक्रियाओं की क्षमता है।

क्या कॉस्मिक किरण मानव शरीर पर प्रभावी हैं?

यदि हाँ, तो किस प्रकार और क्या इस प्रभाव को विज्ञान के क्षेत्र में स्वीकारा जा चुका है?

कॉस्मिक किरणों के संबंध में अभी भी चर्चाएँ हुई; उनसे यह स्पष्ट है कि ऊर्जा-पूंज से मानव घिरा हुआ है और तब क्या यह संभव है कि इतने सशक्त ऊर्जा-पूंज से घिरा हुआ मानव शरीर स्वयं को इन किरणों के प्रभाव से अछूता रख सके। वस्तुतः ये किरणें मानव शरीर पर पूर्णतः प्रभावी होती है। जब साधारण एलेक्ट्रोमैग्नेटिक वेब से संचालित 'रेडिएशन वायोलॉजी' हमें चमत्कृत करता है तो उससे सशक्त ऊर्जा-पूंज के प्रभाव का सहज ही अनुमान लगाया जा सकता है। लेकिन दुःखद तथ्य यह है कि हमारा आधुनिक जीवन-विज्ञान इसके महत्वपूर्ण प्रभाव को स्वीकार करने के बावजूद इसकी अनदेखी कर रहा है। इस प्रकार की उपेक्षा आधुनिक जीवन-विज्ञान की ''हिपोक्रेसी'' का ही परिणाम है।

हालाँकि कॉस्मिक किरणों पर विज्ञान की शाखा ऐस्ट्रोफिजिक्स के तहत अनेक शोध हुए हैं और हो रहे हैं परन्तु जीवन विज्ञान के दरवाजों को कॉस्मिक किरणों के लिए प्रायः बन्द ही कहा जा सकता है। भारत में ही प्रो0 पुष्पा श्रीवास्तव ने गुब्बारा उड़ान द्वारा शैवाल पर कॉस्मिक किरणों के प्रभाव, विभिन्न अवधि में मूल्यांकित एवं चित्रांकित किए, जिनसे मूलतः लाभकारी, विध्वंशकारी तथा उत्परिवर्तित परिणाम उपलब्ध हुए। आशा है कि अगले दशकों में इस पर तेजी से शोध प्रारंभ होंगे, जो मानवता के लिए सर्वथा लाभकारी होंगे और इससे जीवन विज्ञान को तरुणावस्था प्राप्त हो सकेगी।

ऐस्ट्रोजेनेटिक सोसायटी, दरभंगा के शोध-कार्यों से परिलक्षित है कि मानव-शरीर पर इन किरणों का प्रभाव तीन सतहों अर्थात् शारीरिक, आत्मिक एवं अनात्मिक पर पड़ता है। इसे वैज्ञानिक भाषा में ''बॉडी-फिजियोलॉजी' तथा ''जेनेटिक-मैटेरियल'' पर विभिन्न रूप से पड़नेवाला प्रभाव कहा जा सकता है। किरणें शारीरिक प्रक्रियाओं के लिए उचित वातावरण प्रस्तुत करती है। आनुवंशिकी इकाई पर इसके प्रभाव के कारण प्रोटीन बनने की क्रिया के लिए ऊर्जा प्राप्त होती है। चूँकि किरणों का प्रभाव कोण, शक्ति एवं अर्न्तप्रतिक्रिया पर निर्भर करता है, अतएव शारीरिक क्रियाओं के वातावरण में दोष हार्मोन, न्यूरोन, इन्जाइम्स के सृजन में भिन्नता, विभिन्न जीन्स पर इनके विभिन्न प्रभाव के फलस्वरूप होता है। अर्थात् जीन से लेकर विभिन्न शारीरिक क्रियाओं में समाविष्ट ऊर्जा

का उद्गम इसी वाह्य ऊर्जा-स्रोत से होता है। वस्तुतः शरीर-संचालन की प्रक्रिया में जिन विभिन्न ऊर्जाओं की भूमिका वैज्ञानिकों द्वारा स्वीकार की गई है, उन सभी ऊर्जाओं का एकमात्र स्रोत वह किरण-पूंज ही है। इन किरणों का प्रभाव जाइगोट-निर्माण और उसके विकास से मृत्यु पर्यन्त पड़ता है। मानव-शरीर के विकास क्रम में इन किरणों के साथ ही प्राणी के आनुवांशिकी गुणों का भी महत्वपूर्ण योगदान होता है। किरणों के प्रभाव में उनके कोण (व्यक्ति विशेष के लिए), शक्ति अन्र्तप्रतिक्रिया एवं ग्रहों से उनकी प्रतिक्रिया मुख्य हैं। प्रत्येक व्यक्ति विशेष का जन्म खास अक्षांश-देशान्तर पर खास समय में होने पर आकाशखंडों से किरणों का जो कोण बनता है, वह जीवन पर्यन्त प्रभावी रहता है। पृथ्वी पर स्थान परिवर्तन से अन्तर अति सूक्ष्म होता है क्योंकि किरणों की उद्गम स्थान की तूलना में यह दूरी नगण्य रहती है।

क्या कॉस्मिक किरणें ही रोगों का मूल कारण है?

आधुनिक विज्ञान के अनुसार अनेक रोगों का कारण बैक्टीरिया, वाइरस आदि समझा जाता है, तो फिर कॉस्मिक किरणें किस प्रकार रोगों का कारण है?

वस्तुतः विज्ञान द्वारा प्रस्तुत रोगों के कारण का स्पष्टीकरण काफी अस्पष्ट एवं भ्रामक प्रतीत होता है। विशेष अवलोकन एवं गहन मंथन से ये वैज्ञानिक बातें कुछ प्रश्न-चिह्न छोड़ती है। विज्ञान के द्वारा बतलाए गए कारणों से मूलतः आहार-व्यवहार को आधार मान लिया जाता है लेकिन जब उन आहारों से वंचित जन समुदाय इन रोगों से पीड़ित होकर सामने आता है तब ऐसी स्थितियों में रोगों के कारणों के संबंध में चिकित्सक या तो चुप्पी साध लेते हैं अथवा भ्रामक उत्तर देते हैं। उदाहरणार्थ– एक शिशु का कैंसर रोग से ग्रसित होना। वस्तुस्थिति तो यह है कि आधुनिक मेडिकल साइन्स करीब एक हजार रोगों के कारणों को अज्ञात बताता है। अतः रोगों के सभी ज्ञात एवं अज्ञात कारण माध्यमिक हैं। प्राथमिक कारण तो उससे सर्वथा भिन्न विज्ञान का मूल आधार कॉस्मिक किरणें हैं जो शारीरिक क्रियाओं को संचालित करती है। इसके साथ ही विशेष प्रकार की दूषित किरणें तथा कॉस्मिक किरणों के शीतोष्ण प्रभाव शारीरिक अन्तःक्रियाओं में परिवर्तन लाते हैं, जो शरीर के सामान्य संचालन को बाधित करते हुए कभी-कभी बाहरी प्रभावों (बैक्टीरिया, वाइरस आदि) का भी आधार बन जाते हैं। ऐसी अवस्था तब उत्पन्न होती है जब गुणकारी किरणें भी क्षीण या दूषित किरणों से प्रभावित हो जाती है।

ऐस्ट्रोजेनेटिक शोध के क्रम में यह स्पष्ट रूप से देखा गया है कि बैक्टीरिया, वाइरस से हमारे शरीर पर स्वतः प्रभाव नहीं पड़ता है अपितु हमारा शरीर किरणों के प्रभाव से, उन्हें स्वयं आकर्षित करता है। इस प्रकार किरणें शारीरिक क्रियाओं में विषमता उत्पन्न कर इनके लिए स्थान सृजित करती है जो उनकी अवधि पर्यन्त रोग उत्पन्न करते हैं। उदाहरणार्थ; जौन्डिस और निमोनियाँ में एन्टीभाइरस या एन्टी बैक्टीरियल दवा का प्रयोग

न कर मात्र किरणों में परिवर्तन लाने से ही रोग-निवारण संभव हो जाता है। इस प्रकार अन्य रोगों में तो कॉस्मिक किरणों को ही मूल कारण सहजता से माना जा सकता है।

इन सभी तथ्यों के आलोक में ऐस्ट्रोजेनेटिक सोसायटी रोगियों की निःशुल्क चिकित्सा के साथ-साथ विज्ञान की नवीनतम शाखा ऐस्ट्रोजेनेटिक्स अर्थात् खगोल-अनुवांशिकी की निःशुल्क शिक्षा भी प्रदान करती है, जिससे जन-समुदाय के लिए यह विज्ञान रोग-उन्मूलन की दिशा में उत्तरोत्तर लाभकारी सिद्ध हो सके।

क्या वातावरण से इसका सम्बन्ध है?

पृथ्वी पर रासायनिक विकास की प्रारंभिक अवस्था में कॉस्मिक किरणों एवं निर्जीव की प्रतिक्रिया से विभिन्न प्रकार के गैसों की समाहार के रूप में एक ऐसे वातावरण का निर्माण हुआ जिसमें जैविक विकास की प्रक्रिया का आरंभ हुआ फिर इन कॉस्मिक किरणों की जीव-निर्जीव के साथ की प्रतिक्रिया की शुरुआत हुई जिससे वातावरण चक्रवत व्याप्त हो सका। उदाहरणार्थ प्रकाश-संश्लेषण विधि को ही लिया जाए जिससे जीवन का आधार वायु का सृजन होता है। इस विधि में भी जीवित क्लोरोप्लास्ट, निर्जीव जल, कार्बनडाइआक्साइड की सहायता से सौर-ऊर्जा की उपस्थिति में वायु (आक्सीजन) सृजित करता है किन्तु क्रिया पूर्ण तब होती है जब कॉस्मिक किरणों का बचा हुआ भाग ‘‘डार्क-फेज’’ में प्रभावी होता है। अतः वातावरण के इस मूल पहलू के सृजन में कॉस्मिक किरण का पूर्ण योगदान प्रतीत होता है।

स्पष्ट रूप से कॉस्मिक किरणें पौधों के विकास में मुख्य भूमिका निभाते हैं। ये पौध ाों से ऑक्सीजन बनने की प्रक्रिया में प्रभावी होकर अन्य जीवों के लिए प्राण-वायु सृजित करते हैं जिसके फलस्वरूप कार्बन डाइआक्साइड, नाइट्रोजन तथा अन्य गैस सृजित होकर मिश्रित वायु का निर्माण करती है।

वातावरण का दूसरा मुख्य पहलू जल के सृजन का मुख्य आधार भी किरणें ही प्रतीत होती है;

क्योंकि भूगर्भ के सृजन के समय पर रासायनिक पदार्थ एवं किरणें उपलब्ध थी फिर जल का निर्माण हुआ। अतः निर्जीव रसायन एवं किरणों की प्रतिक्रिया से रासायनिक विकास की प्रक्रिया में गैसे (आक्सीजन, हाइड्रोजन आदि) उपलब्ध हुई जिनसे पुनः बाउण्ड फारमेशन में किरणों से ऊर्जा लेकर जल की प्रथम मात्रा निर्मित हुई होगी जो क्रमशः चक्र में जीव एवं वातावरण का आधार बन सका। अतएव कथमपि अतिशयोक्ति नहीं है कि प्राथमिक अवस्था से अद्यपर्यन्त अर्थात् ‘‘उत्पत्ति से विकास तक’’ वातावरण के विभिन्न पहलुओं के कारक के रूप में ये किरणें काम करती हैं।

कॉस्मिक किरणों का रत्नों-उपरत्नों से क्या संबंध हो सकता है? क्या रत्नों-उपरत्नों के उचित प्रयोग से किरणों के दुष्प्रभाव को समाप्त कर रोगों का निवारण संभव है?

कॉस्मिक किरणों का स्रोत 360° अर्थात् पृथ्वी के चारो ओर है जिससे मुख्य रूप से बारह प्रकार की किरणें तीस-तीस डिग्री के आकाश खण्ड से आती है। किरणों में अन्तर्प्रतिक्रिया फिर ग्रहों से प्रतिक्रिया होती है। इस प्रकार किरणें अनन्त किस्मों की हो जाती है लेकिन जब खास प्रकार की किरण खास भूखण्ड के खास रसायन से लगातार प्रतिक्रिया करने लगती है तो वह रासायनिक पदार्थ उस किरण का अवशोषण गुण हजारों वर्ष में प्राप्त कर लेती है एवम् खास केमिकल कम्पाउण्ड के रूप में विकसित हो जाता है। इसके सृजन की विधि प्राकृतिक एवं धीमी होने के कारण ये कम मात्रा में उपलब्ध होते हैं। अतएव ये कीमती पत्थर या रत्न के नाम से प्रसिद्ध होते हैं। कम ही दिनों तक किरणों के प्रभाव से बना हुआ कुछ भिन्न रासायनिक पदार्थ कम क्षमता वाले किरण ग्राही या उपरत्न के रूप में विकसित होते हैं। भूखण्ड पर किरणों का कोण सही होता है लेकिन मानव अलग-अलग स्थानों एवम् समयों में उत्पन्न होकर आकाशखण्ड या किरण से कुछ अच्छा या बुरा जीवन पर्यन्त कोण बनाता है जिसके अनुरूप ही किरणें प्राप्त होती हैं। दूषित कोण से आती हुई किरणें या तो प्राणी को अप्राप्त होते हैं या दुष्प्रभाव का कारण बन जाते हैं। यात्रान्तर्गत किरणें विभिन्न प्रतिक्रियाओं के फलस्वरूप दोष भी ग्रहण करते हैं। इन सबों का परिमार्जन किरण ग्राही पदार्थ (रत्न-उपरत्न) की उचित मात्रा एवं योग से संभव होता है। ऐस्ट्रोजेनेटिक शोधों की शृंखला से स्पष्ट है कि मानव रोग का प्रारंभिक कारण कॉस्मिक किरणें ही है। अप्राप्य या दूषित किरणें या उनके विभिन्न दूषित कोण मानव संचालन की प्रक्रिया (बॉडी फिजियोलोजिकल स्टैटस) को प्रभावित करते हुए अंग-संचालन या विकास को बाधित करता है जिसका अभिलेख उपरोक्त प्रश्न में देखा जा सकता है। इन तथ्यों के आलोक में यह निर्विवाद है कि रत्न एवं उपरत्न का उचित योग एवं मात्रा जो प्रत्येक रोग के लिए भिन्न हो सकती है; रोग निवारण के लिए अत्यधिक प्रभावी है। इसके प्रयोग में दवा की आवश्यकता लगभग नहीं के बराबर होती है। कभी-कभी तो इसकी खास मात्रा में प्रयोग के बाद दवा का प्रयोग बहुत भीषण प्रतिक्रिया दर्शाता है। रोग निवारण प्रक्रिया में रत्नों की भूमिका तो अहम् है ही लेकिन बिना उपरत्न के खास योग एवं मात्रा से पूर्ण रोग निवारण संभव नहीं होता है। अद्यपर्यन्त रोगों के निवारण प्रक्रिया से यह शोध संस्थान इस निष्कर्ष पर पहुँचा है कि मूँगा (इटालियन), गार्नेट, टोपाज, मोती, लाजवर्त्त, गोमेद, टाइगर्स, स्फटिक, ओनेक्स, ए0डी0, पोखराज, पन्ना, हीरा, फिरोजा आदि का विभिन्न योग एवं मात्रा विभिन्न रोगों के निवारण के लिए काफी लाभकारी है। अधिकांश रोग का निवारण उपरत्न से ही संभव हो जाता है लेकिन इसके प्रयोग की प्रक्रिया जटिल एवं पूर्ण वैज्ञानिक होती है। इसका उचित योग एवं मात्रा रोगों से बचाव में कवच का भी कार्य करता है। इससे संचालित प्राणी अनेक रोगों से मुक्ति पाते हैं।

क्या वातावरण संतुलन एवं प्राकृतिक विपदाओं से बचने में ऐस्ट्रोजेनेटिक्स की भूमिका संभव है?

पर्यावरण एवं प्राकृतिक विपदाओं के मूल कारण के रूप में कॉस्मिक किरणों का प्रभाव स्पष्ट होता है क्योंकि एक ओर तो ये स्वस्थ वातावरण उत्पन्न करते हैं लेकिन दूसरी ओर दूषित किरणें अति दुष्प्रभाव छोड़ती है, जिसका परिणाम खास भू-भाग की खास समय में हुई महामारी होती है। खास भूखण्ड पर दूषित किरणों के प्रभाव को भलीभाँति मूल्यांकित किया जा सकता है लेकिन उनका निदान वैज्ञानिक उच्च तकनीक के प्रयोग से ही संभव हो सकता है। दुःखद तथ्य यह है कि हमारे वैज्ञानिक अभी माध्यमिक कारणों मे ही उलझे हुए हैं और प्राथमिक कारण के विषय में कम सोचते हैं जिससे निदान की प्रक्रिया काफी दुर्लभ प्रतीत होती है। ज्ञान की इस मर्माहत करने योग्य स्थिति में निरंतर मानव दूषित किरणों के भीषण-प्रकोपों को झेल रहा है। आधुनिक विज्ञान अभी तक मात्र उस पर मरहम का ही काम कर रहा है, इस भीषण प्रकोपों से विश्व की रक्षा करने की संभावनाओं की ओर से उदासीन बना हुआ है।

एक अणु पर कॉस्मिक किरणों के प्रभाव से उनका एलेक्ट्रॉन लगातार उछलकर एक ऐलेक्ट्रॉनिक बॉडी सृजित करता है जिसे ''औरा'' कहा जाता है। इस प्रतिक्रिया के क्रम में चुम्बकत्व उत्पन्न होता है। इस प्रकार पदार्थ के सूक्ष्मतम कण किरणों के प्रभाव में चुम्बक की तरह व्यवहार करते हैं तो सहज ही अनुमान लगाया जा सकता है कि इतने बड़े भू-खण्ड से कितना अधिक चुम्बकत्व उत्पन्न हुआ होगा। यही चुम्बकत्व वस्तुतः पृथ्वी के गुरुत्वाकर्षण के रूप में परिलक्षित होता है। इसकी संपुष्टि प्रत्येक ग्रह एवम् ब्रह्माण्ड में इसकी उपस्थिति से परिलक्षित है क्योंकि सभी स्थानों पर केमिकल सतह एवम् किरणें निश्चित रूप से व्याप्त है। गुरुत्वाकर्षण की इस सरल प्रक्रिया में जब खास रासायनिक पटल पर खास विध्वंशकारी किरण प्रभावी होती है तो भूचाल की स्थिति उत्पन्न हो जाती है। किरणों का मूल प्रभाव सूक्ष्म तथा कम समयों के लिए होता है अतः वह प्राथमिक कारणों के रूप में स्पष्ट नहीं हो पाता एवं वैज्ञानिक उधेड़बुन में लगे रह जाते हैं। संभव है 'हाई-टेक्नोलॉजी' के प्रयोग से खास प्रभाव की किरणों को रोका जा सके लेकिन विज्ञान की नजरें अभी माध्यमिक क्षेत्र से प्राथमिक क्षेत्र की ओर नहीं उठी है।

जिस प्रकार मानव-शरीर पर दूषित किरणों के प्रभाव से रोग उत्पन्न होते हैं, ठीक उसी तरह दूषित किरण खास भू-भाग पर मानवता या वातावरण को प्रभावित करता है। अतएव यह अतिशयोक्ति नहीं है कि पर्यावरण में अन्य लक्षण प्राथमिक कारण (दूषित किरण) के ही प्रभाव से उत्पन्न होते हैं जिसका समुचित निदान उसमें परिवर्तन से ही संभव है, विज्ञान की दृष्टि उधर जाने भर की देर है। ध्यानाकर्षण मात्र से ही मूल रूप में इन कारणों का परिमार्जन संभव हो सकेगा।

क्या इन किरणों का रंग-विशेष से भी कोई संबंध है?

सूर्य किरणों से उत्पन्न स्पेक्ट्रम से ज्ञात होता है कि उसमें सात रंग होते हैं किन्तु ऐस्ट्रोजेनेटिक शोध से ज्ञात है कि ये वस्तुतः नौ रंग की है जिनमें सफेद तथा मरकरी रंग भी शामिल है। शोध के उपलब्ध परिणामों से स्पष्ट है कि यह स्पेक्ट्रम कॉस्मिक किरणों के मिश्रित रूप को प्रदर्शित करता है जिसका मात्र एक भाग (लाल) सौर ऊर्जा है। अन्य भागों का उद्गम स्थान आकाश है। प्रथम एवं आठवाँ खण्ड नारंगी रंगों की किरणें देता है, दूसरा एवं सातवाँ भाग सफेद, तीसरा एवं छठा खण्ड हरा, चौथा मरकरी, पाँचवा लाल, नौवाँ तथा बारहवाँ पीला, दसवाँ नीला, एवं ग्यारहवाँ आसमानी रंगों का कारक होता है। भूमध्य रेखा के केन्द्र-बिन्दु पर खड़े काल्पनिक रंग के दोनों ओर 15° अर्थात् कुल 30° का प्रथम आकाशखण्ड है। इसके बाद का 30° का दूसरा खण्ड, तीसरा 30° का तीसरा खण्ड और इसी प्रकार ...।

रंगों में इस मूल गुण के कारण ही रंग उपचार 5-10% तक लाभकारी होते हैं। रत्न-उपरत्न में भी रंगों का अत्यधिक महत्व है क्योंकि ये अलग-अलग रंग के किरणों द्वारा ही सृजित होती है।

क्या इन किरणों के उपयोग से मानव-शरीर बिना किसी दवा के सौर-यंत्रों की भाँति संचालित किया जा सकता है?

ऐस्ट्रोजेनेटिक शोध के फलस्वरूप यह ज्ञात है कि उचित शक्ति, कोण एवं अन्तर्प्रक्रिया वाली किरणें मानव विकास के लिए पर्याप्त है। उसकी कमी या दुष्प्रभाव ही विभिन्न रोगों का कारण है। इन कारणों के निदान में प्रकृति में उत्पन्न रत्न और उपरत्न पूर्णतया सक्षम होते हैं। उचित योग तथा मात्रा से लगातार संचालित मानव को कभी भी दवा की आवश्यकता नहीं पड़ सकती है। एकाएक भीषण रोग से ग्रसित अवस्था में 5% रोगों में ही दवा की कुछ आवश्यकता तात्कालिक निदान के लिए होती है क्योंकि रत्न-उपरत्न पूर्ण प्रभाव के लिए कभी-कभी कुछ समय ले लेते हैं। कीटाणु वाले रोगों की अवस्था में भी कभी-कभी (1%) दवा की आवश्यकता तात्कालिक रूप में होती है। वस्तुतः 95% रोगों में रत्न-उपरत्न निराकरण पूर्ण सफल होते हैं। मानवीय भूलों के कारण जब रोग विकराल स्थिति प्राप्त कर लेता है तो कुछ आवश्यक शल्य चिकित्सा करवाने पड़ते हैं। कुछ शल्य चिकित्सा इस विधि से भी संभव हुआ है।

दवाएँ कैसे लाभकारी होती है?

विभिन्न प्रकार की दवाएँ अपनी रासायनिक आण्विक शक्ति द्वारा अंग पर पड़नेवाले विध्वंशकारी किरणों के प्रभाव को अपनी क्षमता से खास अवधि तक रोकती है। कुछ

दवाएँ अन्य शारीरिक क्रियाओं में क्षमता उत्पन्न कर रोग मुक्त करती है। यदि किरणों की भीषणता एवम् अवधि प्रचुर होती तो दवा का प्रभाव खत्म होते ही रोग पुनः उत्पन्न हो जाता है या रोग मुक्ति दवा से संभव ही नहीं होती है। ऐसी स्थिति में चिकित्सक अन्य आयामों का सहारा लेते हैं यथा– आहार, व्यवहार आदि। इस अवस्था में रोगी को तात्कालिक राहत मिलती है लेकिन वह रोग मुक्त नहीं हो पाता है। इसका मूल कारण किरणों का दुष्प्रभाव ही है। रत्न-उपरत्न किरणों के प्रभाव को परिवर्तित करते हैं एवं अन्य शारीरिक प्रक्रियाओं में दृढ़ता लाते हैं। इस प्रकार रत्न-उपरत्न प्राथमिक कारण (किरणों) की स्थितियों को लगातार परिवर्तित कर रोगी को रोगमुक्त करते हैं। चूँकि इनकी मात्रा शरीर के बाहर से ही प्रभावी होती है अतः इसका प्रभाव दवा की तरह समाप्त नहीं होता वरन् लगातार रोग-निदान में सक्रिय होता है।

क्या कॉस्मिक किरणें मानवीय अनुवांशिकता को प्रभावित करती है?

जैसा कि पहले कहा गया है, कि कॉस्मिक किरणें तीन सतहों में प्रभावी होती है। प्रथम दो सतह शारीरिक संस्थान (बॉडी फिजियोलॉजी) से संबंधित है जबकि तीसरी सतह अनुवांशिक गुण (जेनेटिक मैटिरियल) से संबंधित हैं। मूल रूप से जाइगोट निर्माण की प्रक्रिया से लेकर जाइगोट के जेनेटिक मैटेरियल के प्रत्येक जीन को प्रभावित करते हुए शारीरिक विकास एवं उसके दैनिक कार्यकलाप से मृत्यु पर्यन्त प्रभावी होती है। यह तथ्य पीड़ितों के कॉस्मोग्राफ द्वारा स्पष्ट रूप से समझा जा सकता है। अनुवांशिकता के नियमों एवं अनुवांशिकी प्रक्रियाओं (लिंकेज, क्रॉसिंग ओवर, इन्टरेक्शन) के लिए ये किरणें उत्तरदायी है जो प्रायः ऊर्जा पूर्ति के द्वारा उत्पत्ति, विकास एवं प्रजनन से मृत्यु तक प्रभावी होती है। यदा-कदा उत्परिवर्तन की विभीषिका भी सामने आती है। कॉस्मोग्राफ में सारी बातें विभिन्न पीड़ितों में कॉफी रोचक एवं सूक्ष्मतम ढंग से प्रस्तुत प्रतीत होता हैं। चूँकि माता-पिता तथा फिर उनकी संतान में अनुवांशिकी उल्लेख प्रस्तुत होता है अतः कॉस्मोग्राफ का वैज्ञानिक आधार पुष्ट हो जाता है।

कॉस्मोग्राफ प्रत्येक प्राणी के जन्म स्थान अर्थात् विशेष अक्षांश-देशान्तर पर तत्क्षण विभिन्न आकाश खण्ड से आती कॉस्मिक किरणों तथा उनसे प्रतिक्रिया करनेवाले ग्रहों के स्थान का बोध कराता है। इससे उस प्राणी के प्रकृति द्वारा संचालन की शुरुआत में कॉस्मिक किरणों की अन्तर्प्रतिक्रिया, ग्रहों के साथ प्रतिक्रिया, किरणों के कोण तथा किरणों की शक्ति आदि का स्पष्ट लेखा-चित्र प्रस्तुत होता है। इसके आधार पर किरणों के भविष्य में पड़नेवाले प्रभावों का भी मूल्यांकन किया जाता है क्योंकि ग्रहों की गति एवं दूरी नियमित एवम् निर्धारित होती है।

क्या जेनेटिक इंजीनियरिंग को इससे लाभ होगा?

वस्तुतः जेनेटिक इंजीनियरिंग जीन प्रत्यारोपण पद्धति पर आधारित है क्योंकि आधुनिक वैज्ञानिक तकनीक द्वारा जीन के कार्यक्षेत्र में हस्तक्षेप संभव नहीं है। लेकिन ऐस्ट्रोजेनेटिक तकनीकों द्वारा जीन के कार्यक्षेत्र में हस्तक्षेप संभव होता है जिससे बिना प्रत्यारोपण के खास हद तक जेनेटिक इंजीनियरिंग का उद्देश्य पूरा हो जाता है। इस प्रक्रिया में वस्तुतः रत्न प्रभावी होते हैं। उदाहरणस्वरूप इन्सूलीन की कमी वाले रोगी को इन्सूलीन निर्माण चालू करना, थायराइड की गड़बड़ी वाले रोगी को उससे मुक्ति दिलाना; इसी प्रकार मेंटल रिटार्डेशन एवम् कई अन्य अनुवांशिक गुणों को विकसित कर यह आधुनिक जेनेटिक इंजीनियरिंग के क्षेत्र की दृष्टि से काफी अधिक उपयोगी सिद्ध हुआ है।

क्या सारांश में यह कहना उचित होगा कि ये किरणें प्राकृतिक जीव-निर्जीव प्रक्रियाओं का मूल कारक है?

मानव का दैनिक जीवन जीव से जीव फिर निर्जीव से प्रतिक्रिया भर सीमित है। इसी क्षेत्र के तहत सारा जीवन काल सम्पूर्ण होता है। यह प्राकृतिक खेल विशेष मैकेनिज्म पर आधारित है; जिसमें 'औरा' निर्माण मुख्य है। किरणों के प्रभाव से निर्जीव तथा जीव के अणुओं के एलेक्ट्रॉन्स में उछाल के कारण एक एलेक्ट्रॉनिक ढाँचा उत्पन्न होता है जिसे 'औरा' कहा जाता है। जीवत के औरा में विभिन्न रंग होते हैं लेकिन निर्जीव में यह मात्र हल्के पीले रंग का होता है। लेकिन दोनों ही प्रकार के औरा में आकर्षण बल होता है जो किरण के प्रभाव के अनुसार वातावरण करते हैं। दूषित किरणों के संचालन के कारण विषैले तत्वों (वाइरस, बैक्टीरिया आदि) से आकर्षण कभी-कभी काफी भयानक परिणाम उपस्थित करता है। वास्तविक रूप में जीव-निर्जीव प्रक्रिया को ही सर्वसाधारण द्वारा भाग्य कहा जाता है, जिसका संचालन वैज्ञानिक रूप में प्रकृत औरा निर्माण द्वारा करती है और इसका मूल स्रोत कॉस्मिक किरणें हैं। ज्ञातव्य है कि औरा को चित्रांकित करने की वैज्ञानिक तकनीक भी विकसित हो चुकी है जिसे 'किरलन' फोटोग्राफी कहा जाता है। यह मूलतः क्राइम डिटेक्शन के लिए विकसित हुआ।

क्या जीव-निर्जीव के भेद और उनके पारस्परिक संबंध ऐस्ट्रोजेनेटिक्स द्वारा व्याख्यायित किए जा सकते हैं और मानव संचालन में कॉस्मिक किरणों की पूर्ण भूमिका होती है?

पूर्व के प्रश्नोत्तर में कहा जा चुका है कि निर्जीव एवम् किरणों के प्रभाव से जैविक विकास के लिए वातावरण प्रस्तुत हुआ जिसमें जीव उत्पन्न हुए और जो जीवन पर्यन्त

कॉस्मिक किरणों से ऊर्जा प्राप्त कर किरणों की विभीषिका द्वारा अधोगति प्राप्त करते हुए निर्जीव की श्रेणी में पंक्तिबद्ध हो जाते हैं। यही क्रम आध्यात्मिक रूप से पुनर्जन्म या वैज्ञानिक रूप से जीव-निर्जीव प्रक्रिया कहलाता है। प्रायः यह चक्रवत चलायमान है। आज का वृक्ष कालान्तर में हीरा का रूप भी ग्रहण कर सकता है। परिवर्तन की प्रत्येक अवस्था में किरणों का योगदान होता है लेकिन यह स्पष्ट है कि अति प्रभावशाली कॉस्मिक किरणें ही जैविक प्रक्रिया के लिए पूर्ण नहीं है क्योंकि प्रकृति की सारी यांत्रिकी पदार्थ-ऊर्जा परिवर्तन पर आधारित होती है। अतः कॉस्मिक किरण मात्र ऊर्जा है और पदार्थ के रूप में जेनेटिक मैटेरियल होते हैं। इन्हीं दोनों (पदार्थ एवं ऊर्जा) की प्रतिक्रिया से जीव-संचालन होता है और इसी पर आधारित है हमारा नया विज्ञान ''ऐस्ट्रोजेनेटिक्स'' अर्थात् जैविक शरीर या अनुवांशिक गुणों पर अंतरिक्ष किरणों के प्रभाव को वैज्ञानिक धरातल पर व्याख्यायित करनेवाला शास्त्र...।

मानव शरीर का रक्षा-कवच : रत्न और उपरत्न

विक्टर हेस द्वारा प्राप्त अति शक्तिशाली ऊर्जा स्रोत, कॉस्मिक किरणों के रूप में प्रतिपादित हुआ। उनके स्रोत सूर्य, तारे एवं नीहारिकाएँ मानी गई। ऐसा प्रतीत हुआ कि इन किरणों ने पृथ्वी (जड़-चेतन) के विकास में योगदान दिया होगा। ये शक्तिशाली किरणें विभिन्न ऊर्जा क्षमता (1 से 1 लाख वेब) की पायी गई। आधुनिक विज्ञान की ''ऐस्ट्रोफिजिक्स'' शाखा द्वारा इनके भौतिक गुणों पर शोध प्रारम्भ हुआ। जीव व निर्जीव पर इनकी अर्न्तप्रक्रिया अवश्यंभावी प्रतीत हुई। जैविक विकास से पूर्व रासायनिक विकास अनुमानित है। अतएव किरणों का योगदान रासायनिक विकास में मुख्य रहा होगा। खास किरणें हजारों वर्ष तक खास भूखण्ड पर लगातार प्रतिक्रिया के फलस्वरूप कुछ विशेष रासायनिक पदार्थ, जल एवम् स्थल पर निर्मित करती रही होगी? इस प्रकार उनकी खास रासायनिक संरचना के साथ-साथ किरण अवशोषण एवम् परिमार्जन गुण भी विकसित हो गये होंगे? किरणों की क्षमता (पावर) एवम् प्रतिक्रिया की अवधि, इन दो आधारों पर रत्न-उपरत्न का सृजन हुआ होगा? जो आज अपने रासायनिक गुणों एवम् ऊर्जा अवशोषण-परिमार्जन क्षमता को दर्शाते हुए प्रतीत होते हैं।

रासायनिक विकास के साथ या उसके कुछ बाद जैविक विकास की प्रक्रिया में भी इन किरणों का महत्वपूर्ण योगदान रहा होगा? जो रत्न-उपरत्नों की मात्रा एवं खास योग का मानव-शरीर पर पड़नेवाले प्रभाव से परिलक्षित होता है। हजारों प्रायोगिकी के आधार पर यह स्पष्ट होता है कि कॉस्मिक किरणें मानव संचालन एवम् विकास में पूर्ण योगदान देती है। वैज्ञानिक तथ्य के अनुसार ऊर्जा एवम् पदार्थ की प्रतिक्रिया से सारे जैविक कार्य सम्पन्न होते हैं, तो यह सशक्त ऊर्जा स्रोत अनुवांशिक इकाई (जीन्स) पर प्रभावी होकर

प्राणी-संरचना में मुख्य योगदान देता है। किरणों में ऊर्जा-भिन्नता के साथ-साथ स्वभाव भिन्नता-ऊष्ण, शीत, जल, कफ आदि भी प्रामाणिक हैं। कुछ किरणें सुप्रभावी होती है, जिनसे प्राणी-विकास स्वभाविक एवम् सम्यक रूप में होता है। इसके विपरीत कुछ किरणें कुप्रभावी होती है, जो प्राणी विकास में गतिरोध या अस्वाभाविकता उत्पन्न करती है। रत्न और उपरत्न एक ओर जहाँ ऊर्जा परिमार्जन क्षमता द्वारा किरणों के कुप्रभाव को नष्ट करते हैं, वहीं दूसरी ओर ऊर्जा-अवशोषण क्षमता द्वारा सुप्रभावी किरणों का अवशोषण करके मानव-शरीर के लिए रक्षा-कवच की भूमिका का निर्वाह करते हैं। प्राणी-विकास में अस्वाभाविकताओं या प्राणी के अवरोधित विकास का एक कारण सूक्ष्म अनुवांशिक इकाई (जीन्स) की अनुपलब्धि (मिसडिवीजन) या उसकी अनुवांशिक क्षमता है, जो रत्न-उपरत्नों के प्रभाव क्षेत्र से परे है।

प्राणी विकास जाइगोट-निर्माण से प्रारम्भ होता है। इस प्रक्रिया में माता और पिता के गुणों का मेल होता है। इस अनुवांशिक सम्मिश्रण की घड़ी में ऊर्जा-स्रोत (कॉस्मिक किरणों) की स्थिति के आधार पर भावी ऊर्जा-स्रोत के नियमित गुणों का ज्ञान सांख्यिकी द्वारा प्राप्त किया जाता है। सुप्रभावी किरणों से समुचित विकास एवं लाभ प्राप्त होते हैं। इसके विपरीत कुप्रभावी किरणें अवरोधित विकास और जैविक रोग को जन्म देती है। वैसे तो कई ब्रह्माण्ड (गैलेक्सी) की चर्चा आती है, लेकिन हमारा ब्रह्माण्ड पृथ्वी का धारक है, जिसमें कई सौरमंडल, असंख्य तारे एवं नीहारिकाओं के जगमगाते पटल हैं जो स्वयं को गुरुत्वाकर्षण परिधि द्वारा प्रायः अन्य सौरमंडल से पृथक करते हुए स्वतंत्र अस्तित्व स्थापित करते हैं। अतः हमारी पृथ्वी चारो ओर (360°) से किरण पूंज से घिरी हुई है। ये किरणें आकाश खण्ड अर्थात् पृथ्वी से करोड़ों मील दूर से आती है। उनमें से कुछ ओजोन सतह द्वारा अवशोषित तथा पृथ्वी के गुरुत्वाकर्षण बल द्वारा विकर्षित हो जाती है। फिर कुछ किरणें पृथ्वी के वातावरण में प्रवेश करती है। किरणों का प्रत्येक कण पृथ्वी के वातावरण के साथ तीव्र रूप में प्रतिक्रिया कर किरणों की वर्षा करती है। इस प्रकार किरणें पृथ्वी के कण-कण में व्याप्त हो जाती है। फिर उससे लेशमात्र स्थान भी पृथक नहीं रह जाता है। किरणों के इस यात्रा क्रम में उन्हें विभिन्न ग्रहों से भी परावर्तित होना पड़ता है, जिससे उनके गुणों एवं तापक्रम में परिवर्तन होता है, क्योंकि प्रत्येक ग्रह निश्चित पथ पर चलने वाला विशेष रासायनिक पदार्थों का सृजित रूप है, जिसमें विभिन्न शीतोष्ण गुण भी होते हैं।

ज्योतिष शास्त्र के अनुसार ब्रह्माण्ड मानव संचालन 12 राशियों तथा 27 नक्षत्रों द्वारा करता है। ठीक उसी तरहक आधुनिक शरीर विज्ञान के अनुसार 12 जोड़े क्रीनियल स्नायु और 27 मुख्य स्पाइनल स्नायु द्वारा मस्तिष्क शारीरिक अंग संचालन प्रक्रिया स्थापित करता है, जिसमें पूर्ण अंग क्षमता भी परिलक्षित होती है। अर्थात् प्रथम क्रीनियल स्नायु जिस अंग

से जुड़े होते हैं, प्रथम राशि मेष भी उसी अंग का परिज्ञान स्थापित किये हुए है और इसी प्रकार अन्य भी। इन मौलिक गुणों में समानता के कारण ज्योतिष शास्त्र में अनुपलक्ष्य पूर्ण अंग-विचार जीवन-विज्ञान की सहायता से उपलब्ध हो गये हैं, जिनकी प्रामाणिकता ऐस्ट्रोजेनेटिक सोसायटी, दरभंगा द्वारा सिद्ध की जा चुकी है। सोसायटी द्वारा एतत् संबंधी किये गये प्रयोगों को पुनरावृत्ति किसी भी क्षण, कहीं भी की जा सकती है। ऐस्ट्रोजेनेटिक सोसायटी की एक महत्वपूर्ण उपलब्धि यह भी है कि इसके द्वारा शारीरिक प्रक्रियाओं से उत्पन्न पदार्थ पसीना, मूत्र, रक्त आदि विभिन्न ग्रन्थियाँ, स्नायु-संस्थान आदि के संचालक ऊर्जा-स्रोत भी स्पष्ट किये गए हैं। इस तरह जटिल शारीरिक क्रियाओं के सूक्ष्मतम अध्ययन की सहायता से किरणों द्वारा शरीर-संचालन की पूर्ण प्रक्रिया उपलब्ध हो चुकी है।

यात्रा-क्रम में किरणें सर्वप्रथम अपने 12 प्रकार में अन्तर्प्रतिक्रिया करती है। प्रतिक्रियाओं के क्रम में $12 \times 12 \times 27 \times 27 \times 9 \times 9 = 85, 03, 056$ प्रकार की किरणें प्रत्येक मानव पर विभिन्न कोणों तथा शक्ति से बढ़कर उसका संचालन करती है। इन कारणों में विध्वंशकारी किरणें तो होती ही है, साथ ही कुछ लाभकारी किरणें भी। जन्म स्थान के अक्षांश-देशान्तर एवम् समय के कारण खास कोण से आती हुई विध्वंशकारी हो सकती है, जिनका उल्लेख ज्योतिष-शास्त्र में भी है और जिसे प्रयोग द्वारा सत्यापित किया जा चुका है।

दूषित या विध्वंशकारी किरणें जब मानव शरीर के किसी अंग का संचालन खास समय में खास कोण से करने लगती है तो उस समय रोग-विशेष की उत्पत्ति उस अंग में होती है। ऐसी स्थिति खासकर तब बनती है जब अन्य लाभकारी किरणों का संचालन शक्ति के क्षीण हो जाने से शरीर की प्रतिरोधक प्रक्रिया समाप्त हो जाती है। किरणों की कालावधि एवं गुण के अनुरूप रोग 4 प्रकार के होते हैं :-

1. सामयिक 2. जीर्ण 3. अनुवांशिक और 4. उत्परिवर्तित।

सामयिक एवम् जीर्ण रोग में कभी-कभी कीटाणु कारण के रूप में स्पष्ट लगते हैं तथा इनमें से कुछ रोगों को छुआछूत की बीमारी कहा जाता है। वास्तविक रूप में कीटाणु का दुष्प्रभाव भी शरीर पर खास किरणों के पड़ने के बाद खास प्रकार के प्रोटीन निर्माण अथवा शारीरिक परिवर्तन द्वारा कीटाणु के पोषक तत्व तैयार होने पर ही होता है। अर्थात् मूल कारण इन रोगों में भी किरणें ही होती है। किरणें माता-पिता के मूल गुणों को युग्मक (गेमिट्स) द्वारा संतान में पहुँचती है। अनुवांशिकी की यह क्रिया जन्म समय में पारिवारिक ग्रह-स्थिति से स्पष्ट होती है। किरणों के इसी गुण के आधार पर जीवन-विज्ञान की एक नई शाखा ऐस्ट्रोजेनेटिक की अनुशंसा पूरे विश्व के वैज्ञानिक जगत को ऐस्ट्रोजेनेटिक सोसायटी दरभंगा द्वारा की गयी है। साइन्स कॉउन्सिल, जापान ने अपनी खास वैज्ञानिक समिति के समक्ष ऐस्ट्रोजेनेटिक सोसायटी, दरभंगा के शोधपूर्ण कार्यों को

प्रस्तुत किया। समिति ने इससे पूर्णतः सहमति व्यक्त करते हुए इस दिशा में प्रगति के लिए सोसायटी को प्रोत्साहित किया। भारतीय आयुर्विज्ञान संस्थान, दिल्ली के अनुवांशिक विभाग के अध्यक्ष ने भी इसमें काफी रुचि ली। इसकी पूर्ण सूचना कार्य-कलाप के साथ आई0 सी0 एम0 आर0 को भी भेजी जा चुकी है। उपरोक्त पारिवारिक चक्र में अनुवांशिकता के नियम तथा विभिन्न क्रियाएँ भी परिलक्षित होती है। जैसे– लिंकेज, कपलिंग, रिपल्सन, क्रॉसिग ओवर आदि। इससे यह स्पष्ट हो जाता है कि किरणों का प्रभाव अनुवांशिकता की क्रिया में योगदान देता है। इस प्रकार गुणों के साथ-साथ अनुवांशिक रोग भी अगली पीढ़ी में उत्पन्न होते हैं, जिसका दवा से कई गुणा अधिक निवारण रत्न उपचार द्वारा संभव होता है। गुणों की मौलिक इकाई की अनुपस्थिति के कारण उत्पन्न हुए रोगों में निदान संभव नहीं होता। किरणों के गुण-निरूपण के समय यह स्पष्ट हुआ कि कुछ किरणें खास शक्ति के साथ खास कालावधि में उत्परिवर्तन करती है। जैसे– शनि किरणें अपने पूरे चक्र-काल 19 वर्ष में जो प्रभाव छोड़ती है, उससे उत्पन्न रोग पूरे शरीर में एक विशेष वातावरण और प्रोटीन का निर्माण कर देता है जिसका किसी भी उपचार पद्धति द्वारा उपचार संभव नहीं होता है। इस परिस्थिति में भी रोगी का विशेष कष्ट निदान तथा रोग में परिवर्तन इस शास्त्र द्वारा संभव होता है। पूर्व नियंत्रण द्वारा इसे रोका भी जा सकता है।

उपरोक्त रोगों की कम समय में सुलभतम ढंग से की जाने वाली चिकित्सा विधि मानव-कल्याण के लिए विकसित की गयी है, जो ऐस्ट्रोजेनेटिक सोसायटी की एक महत्वपूर्ण उपलब्धि है। कुछ हद तक रोगों से बचने के उपाय भी इन प्राथमिक पहलुओं के नियंत्रण द्वारा शारीरिक क्षमता को बढ़ाकर किए जाते हैं। यह स्पष्ट है कि प्रत्येक मानव सशक्त ऊर्जा-स्रोत से घिरकर उसके सुप्रभाव एवम् दुष्प्रभाव झेल रहा है। यह चिकित्सा-पद्धति पूर्णतः वैज्ञानिक आधार पर विकसित एक अद्भुत प्रणाली है, जिसमें यात्रा एवं योग किरण ाों की अन्र्तप्रतिक्रिया के लिए अति महत्वपूर्ण है। यह प्रणाली अन्य चिकित्सा प्रणाली से पूर्णतः भिन्न प्रक्रिया प्रस्तुत करती है। क्योंकि अन्य चिकित्सा पद्धतियों में औषधियाँ खिलायी जाती है, लेकिन इस पद्धति के अन्तर्गत (चूँकि बाहरी नियंत्रण से आंतरिक संचालन प्रमाणित है, अतएव) रत्नों एवं उपरत्नों को शरीर पर बाँधकर ही प्रभाव दर्शाया जाता है। ये प्रभाव रोग के अनुसार अतिसूक्ष्म काल में भी दर्शाये जाते हैं। उपचार की प्रक्रिया एक दिन से एक माह या जीर्ण रोगों के लिए दो-तीन माह तक की हो सकती है।

रत्न उपचार के खर्चीले होने के भय से अधिकांश रोगों के लिए इसे उपरत्नों पर आध ाारित किया गया है। 384 तरह के उपरत्नों में सस्ते, सुलभ तथा प्रयोग द्वारा पूर्ण प्रभावी 9 प्रकार के उपरत्न चुने गये हैं। इन उपरत्नों का योग एवं स्थान निर्धारित किया जाता है। साधारणतया इनके दाने प्रयोग में लाये जाते हैं; जो विभिन्न रोगों में विभिन्न मात्राओं एवम् योग में दायीं बाँह, गला एवं बायीं बाँह पर बाँध दिये जाते हैं। निश्चित कालावधि

में इनकी मात्रा बढ़ाई जाती है। इस प्रकार उपरत्नों की विभिन्न मात्राएँ चार-पाँच बार तक बढ़ाकर रोग से मुक्ति दिलायी जाती है। दायीं बाँह के लिए तीन या चार तरह के, गले के लिए दो तरह के एवम् बाएँ बाँह के लिए तीन तरह के उपरत्न होते हैं। विषम परिस्थितियों (कैंसर) आदि में रत्नों का प्रयोग किया जाता है, जो अंगूठी के रूप में उँगलियों में धारण कराये जाते हैं। प्रयोग के आधार पर रोगों का सूत्र बनाया जा चुका है। प्रत्येक रोग में आठ या नौ प्रकार के उपरत्न दिये जाते हैं, जिनकी मात्रा रोग के आधार पर निर्धारित की जाती है। उपचार की आरम्भिक अवधि में रोग लक्षण में वृद्धि देखी जा सकती है, जो स्वयं एक-दो दिनों अपनी पूर्ण अवस्था से नीचे पहुँच जाती है। इसके कारण रोग उत्पन्न करने वाली किरणों को दुष्प्रभावी बनाने वाले कोण हैं। ऐसी अवस्था में रोगी को कष्ट से बचाने के लिए उपरत्न की न्यूनतम मात्रा से ही धीरे-धीरे नियंत्रित किया जाता है। विभिन्न प्रकार के रोगों से ग्रसित रोगी के सारे रोगों का ईलाज इस प्रणाली में एक ही साथ किया जाता है। इस स्थिति में प्रतिक्रिया अति तीव्र होती है। शरीर के सभी प्रभावी अंग प्रत्यक्ष या परोक्ष रूप से ठीक होने लगते हैं। परोक्ष से तात्पर्य है रोग-वृद्धि द्वारा भी। जिन स्थानों पर प्रतिक्रिया हो जाती है, वहाँ रोग निवारण सुनिश्चित हो जाता है। प्रतिक्रिया के लिए सिर्फ उचित मात्रा की आवश्यकता होती है। उपचार काल में किसी विशेष परहेज की आवश्यकता नहीं होती है। लेकिन इन उपरत्नों को शरीर के भीतर अपना प्रभाव स्थापित करने में कम से कम 10 दिनों का वक्त लगता है। अतः मात्राओं को बढ़ाने के लिए कालावधि का उचित निर्धारण आवश्यक होता है। आकस्मिक रोगों में भी यह उपचार अति लाभकारी सिद्ध होता है, लेकिन इसमें मात्राएँ अधिक लगती है। कभी-कभी इस उपचार में रोगी की शल्य चिकित्सा द्वारा रोग मुक्त कराया जाता है, वैसे पूर्ण नियंत्रित रहने पर शल्य चिकित्सा की स्थिति भी नहीं आती है। उल्लेखनीय है कि रत्न चिकित्सा आयुर्वेद में भी की जाती है, किन्तु वहाँ भस्म बनाये जाने से रत्न समाप्त हो जाते हैं, जबकि इस पद्धति में रत्न और उपरत्न सुरक्षित भी रहते हैं और रोग का उपचार भी हो जाता है। बचे हुए रत्न और उपरत्न अन्य रोगियों को चिकित्सा के लिए पूर्ववत प्रभावी होते हैं। कॉस्मिक किरण ों के दीर्घावधि में पड़ने वाले विध्वंशकारी प्रभाव से स्वस्थ व्यक्तियों की भी रक्षा करने के लिए उन्हें रत्नों एवं उपरत्नों की समुचित मात्रा से नियंत्रित किया जाता है। इस प्रकार ये रत्न और उपरत्न कॉस्मिक किरणों से मानव शरीर की रक्षा के लिए अभेद्य रक्षा कवच का काम करते हैं तथा इन किरणों के कुप्रभाव को नगण्य बना देते हैं।

इस नियंत्रण प्रणाली में रोगों का ज्ञान किरणों की स्थिति से होता है। जन्म-समय का परिज्ञान उपस्थित नहीं रहने पर हाथ में किरण पर्वत के अध्ययन से किरणों की स्थिति का ज्ञान होता है। रोगों का पूर्ण वैज्ञानिक अध्ययन (इटियोलॉजी ऑफ पेशेन्ट) एक्स-रे, पैथोलॉजिकल जाँच द्वारा भी किया जाता है और नियंत्रण के बाद पुनः इन सभी जाँचों के द्वारा रोग मुक्ति दर्शायी जाती है।

''ऐस्ट्रोजेनेटिक्स'' प्रकृति द्वारा मानव संचालन प्रणाली का सूक्ष्मतम अध्ययन प्रस्तुत करता है, साथ ही प्रकृति के विशिष्ट पहलुओं जड़-चेतन प्रक्रिया, आकर्षण-विकर्षण, प्राकृतिक विपदाएँ आदि पर भी अध्ययन के द्वार खोलता है। चिकित्सा पद्धति की यह सुलभतम एवम् अति प्रभावशाली प्रतिक्रिया वैज्ञानिक जगत के लिए विशेष रूप से उपयोगी है। आधुनिक तकनीक द्वारा विषम परिस्थिति (अति क्लिष्ट रोगों) में भी आशातीत सफलता प्राप्त की जा सकती है। आधुनिक चिकित्सा विज्ञान रोगों के कारण के संदर्भ में काफी अस्पष्ट प्रतीत होता है।

ऐस्ट्रोजेनेटिक सोसायटी-सचिव प्रतिवेदन

ऐस्ट्रोजेनेटिक सोसायटी, दरभंगा का सचिव पद पर कार्य करने में मुझे आनन्द का अनुभव हो रहा है। यह सोसायटी प्रकृति द्वारा मानवीय संचालन के वैज्ञानिक आधार का अध्ययन करने के साथ-साथ मानवतावादी धर्म की स्थापना चाहता है। ताकि समाज के क्षीण वैभव तथा स्वास्थ्य की पुर्नस्थापना की जा सके। यह सोसायटी समाज में स्थापित अन्य सोसायटी से भिन्न अस्तित्व का है क्योंकि इसका प्रत्यक्ष संबंध मानवीय व्यक्तित्व (शारीरिक/मानसिक) से है। सोसायटी के द्वारा किए गए सभी कार्यों का विवरण देना यहाँ संभव नहीं है फिर भी प्रमुख कार्यों का लेखा-जोखा संक्षेप में प्रस्तुत कर रहा हूँ। :-

शैक्षिक अनुसंधान : सोसायटी विभिन्न वर्गों से जुड़े लोगों को मान्य सिद्धांतों एवम् पाठ्यक्रम के अनुरूप शिक्षा दे रहा है, साथ-साथ अनुसंधान कार्य हो रहे हैं। वर्ष 1990-91 में 20 छात्र/छात्रा अध्ययनरत रहे।

रोग निवारण एवम् निदान : सोसायटी के द्वारा विभिन्न प्रकार के रोगों यथा– कैंसर, दमा, एलर्जी, गैस्टिक, मधुमेह, चर्मरोग, स्त्री जनित रोग, जौन्डिस, किसी भी प्रकार का दर्द, गठिया, बात, विभिन्न प्रकार के ज्वर आदि पर अध्ययन करते हुए निवारण एवम् निदान के लिए चिकित्सा व्यवस्था की गयी जिससे लगभग पाँच हजार से अधिक लोग 1990-91 में लाभान्वित हुए जिनका इटियोलॉजी Ref No. 1/7 से 40/16 तक वर्णित है।

इन कार्यों के अलावा सामाजिक एवम् सांस्कृतिक क्षेत्र में सोसायटी के द्वारा कार्य किए जा रहे हैं। आर्थिक रूप से पंगु यह सोसायटी अनेक कार्य राशि के अभाव में नहीं कर पा रहा है।

संस्थापक डॉ0 सी0 एन0 झा, अग्रज प्रो0 नृपेन्द नाथ झा और डॉ0 विनय कुमार सिंह, अभिन्न मित्र अनिल कुमार झा एवम् अन्य सहयोगियों जैसे नरेश, नीरव, श्रवण कुमार झा के प्रति कृतज्ञ हूँ जो सोसायटी को सर्वोच्च लक्ष्य पर पहुँचाने के लिए प्रतिबद्ध हैं।

आईये, हम एक साथ मिलकर समाज के पुनरुत्थान के लिए मूल्यों का पुनर्निरूपण करें।

मनोज कुमार ''राजेश''

कॉस्मिक किरण

स्वच्छन्द आकाश के जगमगाते चाँद-सितारे।
दिन के सूर्य के रश्मियों के साथ।।
प्रस्तुत करते हैं :-
जीव-निर्जीव संचालन प्रक्रिया का आधार।
तिल-तिल में व्याप्त होकर।।
प्रस्तुत करते हैं :-
आध्यात्मिक प्रक्रिया का आधार।
जीवन-प्रणाली के विभिन्न मार्गों में।
प्रस्तुत करते हैं :-
सामाजिक ज्ञान का आधार।
तो, —?
यही ज्ञान है, यही विज्ञान।
यही वेद है, यही पुराण।।
यही आन है, यही है शान।
यही रोग है, यही निदान।।
फिर रहता नहीं कुछ हाथ में बाँकी।
तो क्यों न करें प्रस्तूत मानवता की झाँकी।।

संगीत कुमार 'आशीष'
कटहलबाड़ी, दरभंगा

काली कान्त झा

एडवोकेट

काली कान्त झा
एडवोकेट
सिविल कोर्ट, दरभंगा

प्रो० (डा०) इन्द्र नाथ झा, पूर्व विभागाध्यक्ष वनस्पति शास्त्र विभाग, ल० ना० मि० मिथिला विश्वविद्यालय, दरभंगा द्वारा संचालित हेल्थ्रो जेनेटीकल संस्था से मैं काली कान्त झा, एडवोकेट पूर्ण परिचित हूँ। वर्ष 1993-94 की बात है मेरी पत्नी सुनीता झा असाधारण विकार रहती थी। प्रति दिन 15-20 पैखाना पतला हुआ करता था। दरभंगा शहर के प्रमुख-2 डाक्टर से इलाज होता था परन्तु कोई विशेष फायदा नहीं दिख पड़ता था। इसी समय में मेरे प्रिय मित्र स्व० उमानन्द मिश्र ज्योतिषाचार्य पटना से मेरी मुलाकात दरभंगा के श्यामा मंदिर परिसर में हुई। उन्होंने मुझे उपरोक्त संस्था के बारे में जानकारी दी एवं कहा कि वे उठकर खड़ा नहीं हो सकते थे एवं उनका इलाज एम्स दिल्ली से हुआ था, परन्तु कोई फायदा नहीं, लेकिन जब से प्रो० (डा०) इन्द्र नाथ झा के द्वारा Stone Therapy करवायें तो चलना ही नहीं, साइकिल भी चलाने लगे। संयोग की बात है कि उसी समय में प्रो० झा के प्रेस कनफेंस में मैं भी मौजूद था। पत्रकार द्वारा पूछे गये प्रश्न एवं प्रो० झा द्वारा दिये गये जानकारी से मैं पूर्ण संतुष्ट हुआ एवं इस निर्णय पर पहुँचा कि यह Stone Therapy तो विज्ञान है। मैं दूसरे दिन उनसे मिला एवं उन्होंने अपना बहुमूल्य समय देकर कृतार्थ किया। मेरी पत्नी का इलाज (Stone therapy) शुरू हुआ। तीन दिन के अन्दर ही अभूतपूर्व फायदा का अनुभव हुआ। एक महीने के इलाज में मेरी पत्नी सुनीता झा पूर्ण स्वस्थ हो गई। आश्चर्य की बात है कि वर्ष 1993-94 से लेकर अभी तक पूर्ण स्वस्थ हैं। कभी कोई दिक्कत होती है भी तो दिये गये Stone के Doze को घटाकर या बढ़ाकर ठीक कर लिया जाता है।

अतः प्रो० झा द्वारा संचालित संस्था परम लोकप्रिय, परोपकारी सुन्दर, शुद्ध, सुव्यवस्थित एवं लोक कल्याणकारी है, खासकर आयहीन वंचित लोगों के लिये तो ये भगवानस्वरूप हैं, क्योंकि यहाँ इलाज मुफ्त में होता है।

मैं प्रो० झा के इस लोकोपकारी कार्य के लिये बहुत बहुत अभिनन्दन एवं उत्साहवर्धन करता हूँ।

Made in the USA
Monee, IL
12 November 2026

a17da784-85a8-484f-84e9-cf45a07eb63eR01